International Conference on Structural Nonlinear Dynamics and Diagnosis (CSNDD 2018)

MATEC Web of Conferences Volume 241 (2018)

Tangier, Morocco
25 - 27 June 2018

Editor:

Mohamed Belhaq

ISBN: 978-1-5108-7595-1

Printed from e-media with permission by:

Curran Associates, Inc.
57 Morehouse Lane
Red Hook, NY 12571

Some format issues inherent in the e-media version may also appear in this print version.

No changes have been made to the content of these proceedings. There may be changes to pagination and minor adjustments for aesthetics.

Printed by Curran Associates, Inc. (2019)

For additional information, please contact EDP Sciences – Web of Conferences at the address below.

EDP Sciences – Web of Conferences
17, Avenue du Hoggar
Parc d'Activité de Courtabœuf
BP 112
F-91944 Les Ulis Cedex A
France

Phone: +33 (0) 1 69 18 75 75
Fax: +33 (0) 1 69 28 84 91

contact-edps@webofconferences.org

Additional copies of this publication are available from:

Curran Associates, Inc.
57 Morehouse Lane
Red Hook, NY 12571 USA
Phone: 845-758-0400
Fax: 845-758-2633
Email: curran@proceedings.com
Web: www.proceedings.com

TABLE OF CONTENTS

Preface

The 4th CSNDD 2018 was organized by the nonlinear dynamic group of University Hassan II-Casablanca in the city of Tangier in June 25-27, 2018. The conference gathered scientists working in nonlinear dynamics and control in different branches of applied mathematics, advanced physics and mechanics to discuss the latest achievements and to exchange ideas in theoretical, numerical and experimental techniques. Focuses were directed toward diverse topics, ranging from dynamical systems theory to different physical and engineering applications. The CSNDD 2018 covered a large field of nonlinear dynamics, including:

- Dynamics of micro- and nano-electro-mechanical systems
- Vibration energy harvesters, nonlinear dampers and vibration absorbers
- Deterministic and stochastic dynamics
- Mechanical, aeronautic, aerospace and naval structures
- Multiphysics problems in micro and nanosystems
- Nonlinear dynamics of rotating machines
- Time delay in nonlinear systems
- Quasi-periodic oscillations in nonlinear systems
- Linear and nonlinear vibrations of periodic and near-periodic lattices
- Nonlinear phenomena of electrical power and energy circuits and systems
- Time-periodic systems: Theory and applications
- Nonlinear vibrations of cables, beams, plates, shells, and innovative structures
- Nonlinear dynamics of aeroelastic and hydroelastic systems
- Analytical and semi-analytical methods in nonlinear dynamics
- Model reduction methods dedicated to low/high frequency system dynamics
- Structural health monitoring and damage detection

The meeting has attracted more than 120 delegates from 29 countries. Twelve keynote lectures were invited and 18 minisymposia organized by leading experts in nonlinear dynamics. More than 90 oral presentations and 40 posters were scheduled.

The organizing committee thinks the Keynote lecturers, the invited speakers, the minisymposia organizers and reviewers for their precious help in maintaining a high-quality scientific program of the conference.

Mohamed Belhaq
CSNDD 2018 Chair

edp sciences

Web of Conferences

Statement of Peer review

In submitting conference proceedings to *Web of Conferences*, the editors of the proceedings certify to the Publisher that

1. They adhere to its **Policy on Publishing Integrity** in order to safeguard good scientific practice in publishing.
2. All articles have been subjected to peer review administered by the proceedings editors.
3. Reviews have been conducted by expert referees, who have been requested to provide unbiased and constructive comments aimed, whenever possible, at improving the work.
4. Proceedings editors have taken all reasonable steps to ensure the quality of the materials they publish and their decision to accept or reject a paper for publication has been based only on the merits of the work and the relevance to the journal.

Title, date and place of the conference

International Conference on Structural Nonlinear Dynamics and Diagnosis (CSNDD 2018)
Tangier 25-27 June, 2018

Proceedings editor(s):

Mohamed Belhaq

Date and editor's signature

November 02, 2018

MATEC Web of Conferences **241**, 01001 (2018) https://doi.org/10.1051/matecconf/201824101001
CSNDD 2018

Dynamic analysis and characterization of a nonlinear bi-stable piezo-magneto-elastic energy harvester

João Victor Ligier Lopes Peterson[1,*], Vinicius Gonçalves Lopes[1,**], and Americo Cunha Jr[1,***]

[1]Universidade do Estado do Rio de Janeiro

Abstract. Energy harvesting is a very promising technology to provide low levels of power for small autonomous systems, which the applicability encompass a very wide range of areas, that spans from micro/nano sensors in engineering to state of art implants in medicine. The present work deals with the analysis and detailed characterization of a nonlinear bi-stable piezo-magneto-elastic energy harvester driven by a periodic external excitation. The dynamical system is studied in depth through bifurcation diagrams and basins of attraction. The level of chaoticity of the dynamical system is accessed very efficiently via the 0-1 test for chaos, which allows mapping the presence of dense regions of chaos without the help of the Lyapunov exponents.

1 Introduction

Energy harvesting technologies are very promising alternatives for electrical supplying of low power systems, such as medical implants, micro electro-mechanical systems (MEMS), nano electro-mechanical systems (NEMS), etc [1, 2]. Such devices converts environmental energy, available for free, into electrical power to be used in the application of interest. Due to the immense possibility of applications, together with the theoretical challenges associated with the analysis of the underlying nonlinear behavior, these dynamical systems have been object of study of several recent works [3–8].

The present paper aims to investigate the nonlinear dynamics of the bi-stable piezo-magneto-elastic energy harvester by Erturk et al. [9]. In particular, it is of interest to map out whether the system response is chaotic or regular, since the nature of the response of the system has great influence on the level of energy possibly recoverable.

The dynamic system of interest is investigated via bifurcation diagrams and basins of attraction, where the latter are computed not in the traditional way, but with the aid of a binary test for chaos detection.

The rest of this paper is organized as follows. The second section introduces the harvester device of interest and the mathematical model used to analyze its dynamic behavior. Bifurcation diagrams are presented in the third section, while an analysis with 0-1 test for chaos and basins of attraction is conducted in the fourth section. Finally, the fifth section, summarizes the paper main conclusions and contributions.

*e-mail: joao.peterson@uerj.br
**e-mail: vinicius.g.lopes@uerj.br
***e-mail: americo@ime.uerj.br

2 Mathematical modeling

2.1 Physical model

A schematic representation of the bi-stable piezo-magneto-elastic energy harvester studied here is depicted on Figure 1. A slim ferromagnetic cantilever beam, exposed to magnetic fields effects, excites the piezo-electric material plates placed on its upper edge. The rigid structure supporting the whole ensemble is subjected to a periodic movement which drives the nonlinear system.

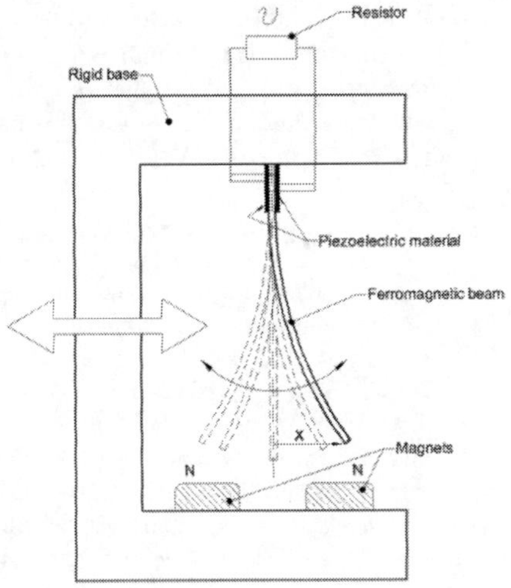

Figure 1. Schematic representation of the bi-stable piezo-magneto-elastic energy harvester by [9].

2.2 Model equations

The dynamic behavior of physical system in Figure 1 evolves according to an evolution law dictated by the pair of equations

$$\ddot{x} + 2\xi\,\dot{x} - 0.5\,x(1 - x^2) - \chi\,v = f\,\cos(\Omega\,t), \quad (1)$$

$$\dot{v} + \lambda\,v + \kappa\,\dot{x} = 0, \quad (2)$$

where χ and κ represents the specific piezoelectric coupling terms, ξ the mechanical damping ratio, λ a time constant reciprocal and, finally, f and Ω represent the excitation amplitude and frequency, respectively. The beam displacement x, its velocity $\dot{x}$ and output voltage v are obtained through the integration of the initial problem value (IPV) defined by Eqs. (1) and (2) and appropriate initial conditions, using the values $\xi = 0.01$, $\chi = 0.05$, $\lambda = 0.05$ and $\kappa = 0.5$. All the aforementioned quantities are dimensionless.

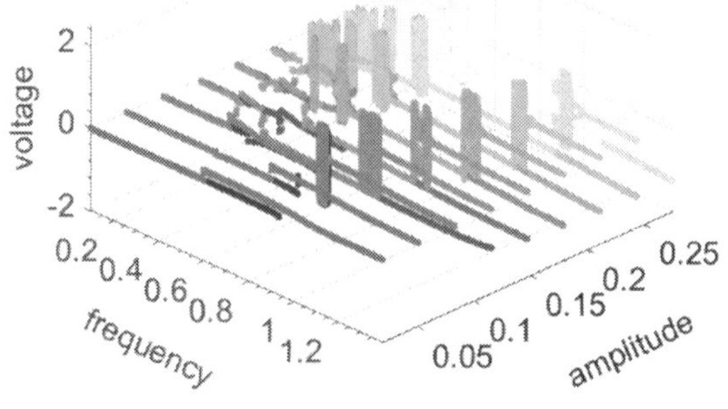

Figure 2. Bifurcation diagrams of the output voltage as function of the excitation frequency, for several values of excitation force amplitude. Forward analysis are presented blue tones, while backward analysis appear in red tones.

3 Bifurcation diagrams

In order to identify if the dynamic behavior of the bistable harvester device is regular or chaotic, this section makes an analysis based on bifurcation diagrams, where the output voltage is considered as an observable of the dynamical system, which is calculated as a function of a certain control parameter (excitation force amplitude f or excitation force frequency Ω).

These diagrams are constructed increasing (forward analysis) or decreasing (backward analysis) the control parameter in a quasi-static way, plotting on the vertical axis the number of periods associated with the current solution. This analysis establishes an interval of interest for the control parameter (f or Ω), which is discretized by an uniform mesh with 1200 points. For all cases under analysis, initial displacement, velocity and voltage are set as $(x_0, \dot{x}_0, v_0) = (1, 0, 0)$. Only the last 10% of the time series are considered in the construction of the diagrams.

3.1 Fixed amplitude, variable frequency

The first analysis considers Ω as the control parameter, observing the system dynamics in the interval $0.1 \leq \Omega \leq 1.4$, for a fixed set of forcing amplitude $f \in \{f_n = 0.019 + (n-1)0.032,\ n = 1, \cdots, 9\}$, as can be seen in Figure 2.

The backward diagram reveals a chaotic dynamic behavior for higher frequencies in the addressed interval and $f \geq 0.051$. Multiple period responses comes up close to the chaotic regions for $f \geq 0.179$. Regions of regular (non-chaotic) dynamics can be seen between chaotic areas.

For both diagrams, forward and backward, a spread region with discontinuities emerges on low frequencies interval, for amplitudes between $f = 0.083$ and 0.211. Regular dynamics can be observed for an entire interval of frequencies only when $f = 0.019$, despite of non-zero values been registered for $\Omega \geq 0.8$ on the forward diagram.

A general analysis suggests that negative regular voltages appears, in the forward diagram, spread on high frequencies interval, and at the central region on both diagrams, specially for $f \geq 0.179$. Positive response predominates for all f interval on higher frequencies values over $\Omega = 0.8$, with some of then also presenting chaotic patterns.

3.2 Fixed frequency, variable amplitude

In the second analysis f is taken as control parameter, observing the system dynamics in the interval $0.05 \leq f \leq 0.3$, for a set of fixed forcing frequencies $\Omega \in \{\Omega_n = 0.1 + (n-1)0.1,\ n = 1, \cdots, 9\}$, as can be seen in Figure 3.

It is possible to see that regular behavior dominates the lower portion of f interval, for frequencies smaller than $\Omega = 0.8$, in both diagrams (forward and backward). Although, the same analysis reveals that voltage reaches zero on both cases, for frequencies $\Omega < 0.5$. Regular negative values, indicating poles inversion on piezoelectric terminals, can be noted on backward diagrams for $\Omega \geq 0.6$, for almost all amplitudes analyzed.

Chaotic dynamics have a strong presence for low frequencies until $\Omega = 0.5$ on both diagrams, even for amplitude from $f \geq 0.119$. Forward diagrams shows chaotic patterns emerging on the middle portion of f interval for frequencies as $\Omega = 0.8$ and $\Omega = 0.9$, while backward diagrams remains regular, presenting a similar profile since $\Omega = 0.6$.

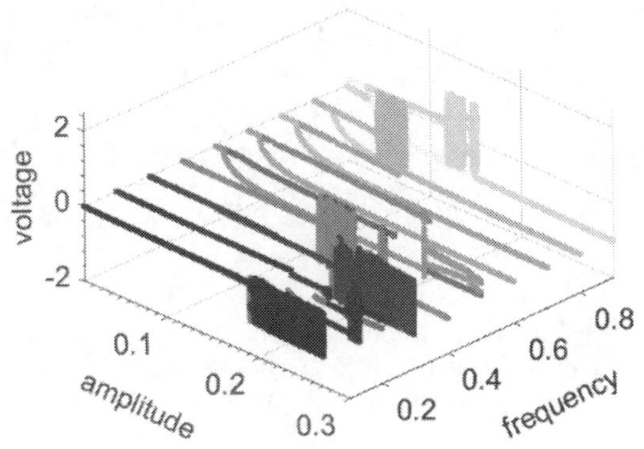

Figure 3. Bifurcation diagrams of the output voltage as function of the excitation force amplitude, for several values of excitation frequency. Forward analysis are presented blue tones, while backward analysis appear in red tones.

Other regular results are obtained for $\Omega = 0.7$, where backward diagrams reveals negative voltages for $f \geq 0.03$, and in the forward diagrams for $f = \geq 0.11$.

The two diagrams reveal the coexistence of stable solutions for the dynamic system of interest, depending on the configuration of parameter adopted for the external excitation.

4 Basins of attraction

Another tool used here to characterize the system dynamic behavior of the bi-stable harvester is the 0-1 test for chaos by Gottwald and Melbourne [10]. This statistical test uses a binary classifier $K \in \{0, 1\}$ to identify the regime of the dynamics — $K = 0$ means regular dynamics, while $K = 1$ stands for chaotic behavior. The classifier is constructed with basis on an extension of the dynamical system to a two-dimensional Euclidean group, that (asymptotically) behaves likes a Brownian motion, in case of chaotic dynamics, and generates limited trajectories for deterministic dynamics. A time series of the dynamical system is projected into an extended space and undergoes a sequence of transformations, to generate the classifier, that depends on a parameter randomly c. The test is repeated hundreds of times, each time with a new value of c, and the final classifier is given by the median of the several samples of K previously generated. Further details about this test can be seen in [10, 11].

Figure 4 shows the projection of a basins of attraction in the displacement vs velocity plane, obtained from the 0-1 classifier, for excitation amplitude and excitation frequency values equal to $f = 0.083$ and $\Omega = 0.8$, respectively. It is seen that chaos and regularity have well-defined regions, with blue representing chaotic responses and red regular ones.

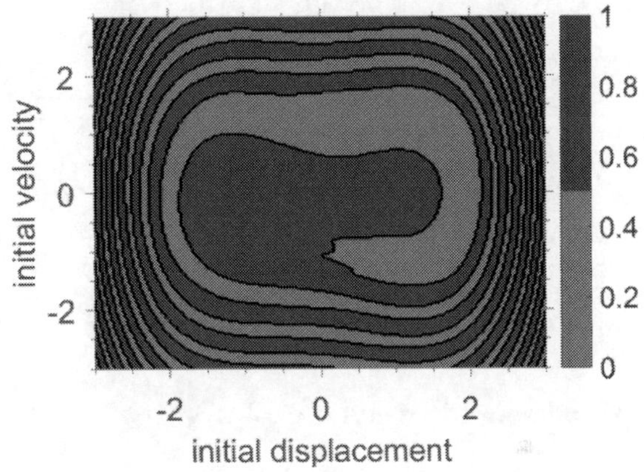

Figure 4. Projection of the basins of attraction in the displacement vs velocity plane, obtained with the 0-1 test for chaos, for $f = 0.083$ and $\Omega = 0.8$.

In Figure 5 it is shown the projection of the basin of attraction in displacement vs velocity plane, for $f = 0.083$ and $\Omega = 0.9$. The results display a behavior very similar to the previous case, although regular responses appear embedded in previous chaotic regions.

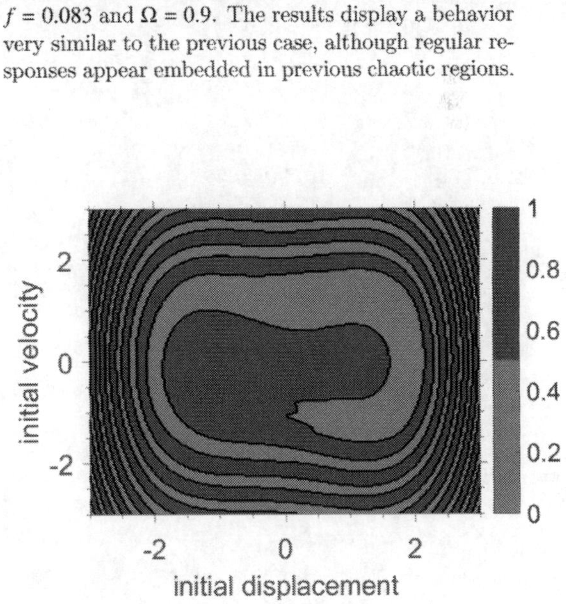

Figure 5. Projection of the basins of attraction in the displacement vs velocity plane, obtained with the 0-1 test for chaos, for $f = 0.083$ and $\Omega = 0.9$.

5 Final remarks

This paper addresses the nonlinear dynamic behavior of a bi-stable piezoelectric energy harvester regarding chaos incidence over system output voltage profile. Bifurcation diagrams and basins of attraction (obtained with aid of 0-1 test for chaos) are employed to distinguish the dynamic behavior of the system between regular and chaotic. The analyzes reveal a very rich dynamic behavior, with the coexistence of stable solutions and dense regions of chaos.

This work is one more in a series where the authors explore in detail the dynamics of the system of interest aiming to better know their physics to optimize the energy recovery process. In a next step, the authors intend to analyze the effect of parameter uncertainties on the system dynamic behavior.

Acknowledgments

The authors are indebted to the Brazilian agencies CNPq, CAPES, and FAPERJ for the financial support given to this research.

References

[1] F. Cottone, H. Vocca and L. Gammaitoni, *Nonlinear Energy Harvesting*, Physical Review Letters, **102**, 080601 DOI 10.1103/PhysRevLett.102.080601 (2009)

[2] P. Spies, M. Pollak and L. Mateu, *Handbook of Energy Harvesting Power Supplies and Applications* Pan Stanford (2015)

[3] A. Pfenniger, A. Stahel, V. M. Koch, D. Obrist and R. Vogel, *Energy harvesting through arterial wall deformation: A FEM approach to fluid-structure interactions and magneto-hydrodynamics*, Applied Mathematical Modelling, **38**, 3325-3338 DOI 10.1016/j.apm.2013.11.051 (2014)

[4] S. Nabavi and L. Zhang, MEMS piezoelectric energy harvester design and optimization based on Genetic Algorithm In: *2016 IEEE International Ultrasonics Symposium (IUS)* DOI 10.1109/ultsym.2016.7728786 (2016)

[5] A. Cunha Jr, *Enhancing the performance of a bi-stable energy harvesting device via cross-entropy method* https://hal.archives-ouvertes.fr/hal-01531845

[6] J. V. L. L. Peterson, V. G. Lopes and A. Cunha Jr, Maximization of the electrical power generated by a piezo-magneto-elastic energy harvesting device In: *XXXVI Congresso Nacional de Matemática Aplicada e Computacional* DOI 10.5540/03.2017.005.01.0200 (2016)

[7] V. G. Lopes, J. V. L. L. Peterson and A. Cunha Jr, Numerical study of parameters influence over the dynamics of a piezo-magneto-elastic energy harvesting device In: *XXXVII Congresso Nacional de Matemática Aplicada e Computacional* DOI 10.5540/03.2018.006.01.0407 (2017)

[8] V. G. Lopes, J. V. L. L. Peterson and A. Cunha Jr, On the nonlinear dynamics of a bi-stable piezoelectric energy harvesting device In: *24th ABCM International Congress of Mechanical Engineering* DOI 10.26678/abcm.cobem2017.cob17-1570 (2017)

[9] A. Erturk, J. Hoffmann and D. J. Inman, *A piezomagnetoelastic structure for broadband vibration energy harvesting*, Applied Physics Letters, **94** 254102 DOI 10.1063/1.3159815 (2009)

[10] G. A. Gottwald and I. Melbourne, The 0-1 Test for Chaos: A review In: C. Skokos, G.A. Gottwald, and J. Laskar (Eds.). *Chaos Detection and Predictability, Springer Lecture Notes in Physics 915* DOI 10.1007/978-3-662-48410-4 (2016)

[11] D. Bernardini and G. Litak, *An overview of 0-1 test for chaos*, Journal of the Brazilian Society of Mechanical Sciences and Engineering, **38** 1433-1450 DOI 10.1007/s40430-015-0453-y (2016)

MATEC Web of Conferences **241**, 01002 (2018)
CSNDD 2018
https://doi.org/10.1051/matecconf/201824101002

Linear flexural natural frequencies and stability analysis of spinning Rayleigh beams: application to clamped-clamped beams.

Mohamed Amine Aouadi, Faouzi Lakrad

Laboratory of Renewable Energy and Dynamics of Systems, Faculty of Sciences Ain Chock, University Hassan II-Casablanca, Morocco

Abstract. In the present paper 3D bending linear free vibrations of spinning Rayleigh beams are investigated. Four linear models, that differ in the linearization process, are studied. A focus on analytical computation of natural frequencies for a broad range of boundary conditions is highlighted. Then, the conditions of occurrence of divergence and flutter instabilities are determined. Finally, a case study consisting of a clamped-clamped Rayleigh beam is studied. It is found that the free vibrations destabilization process depends on the used linearization approach.

1 Introduction

Spinning beams along their longitudinal axes are used in many space and mechanical structures and they are an important part of rotating machines [1] and [2]. Spinning can cause rigid body motions, axial, torsional and 3D flexural vibrations. Moreover, spinning introduces gyroscopic effects, including Coriolis, centrifugal, and Euler forces. This latter force appears when spinning velocity is non-constant.

Aouadi and Lakrad [3] made a literature survey on 3D bending linear models of spinning Rayleigh beams. They focused on four models that differ according to the introduction stage of the linearization process in the computations. They expressed these four models in a condensed model and discussed the results of a double-hinged spinning Rayleigh beam.

The present paper is a continuation of the work [3]. Hence, the main contributions of the present work are :

1. Analytical derivation, in a general framework, of natural frequencies and stability analysis of the condensed model of 3D flexural vibrations of spinning Rayleigh beams [3].
2. Study of clamped-clamped spinning Rayleigh beams.

The present paper is organized as follows: in section 2, the mathematic model of spinning Rayleigh beams introduced in [3] is presented. In section 3, natural frequencies and stability analysis are investigated in general framework. In section 4, the clamped-clamped spinning Rayleigh beam is studied.

2 Formulation of the problem

The mathematical modelling used in the present paper is based mainly on the work [3].

The following assumptions are used

1. The Rayleigh beam theory is used.
2. The beam is straight in its undeformed state.
3. The beam has a uniform cross section, and uniform and isotropic material properties along its length.
4. Cross-sections of the beam have at least two axes of symmetry.
5. The effect of gravity and damping are not taken into account.
6. The spinning axe is the longitudinal direction of the undeformed beam.

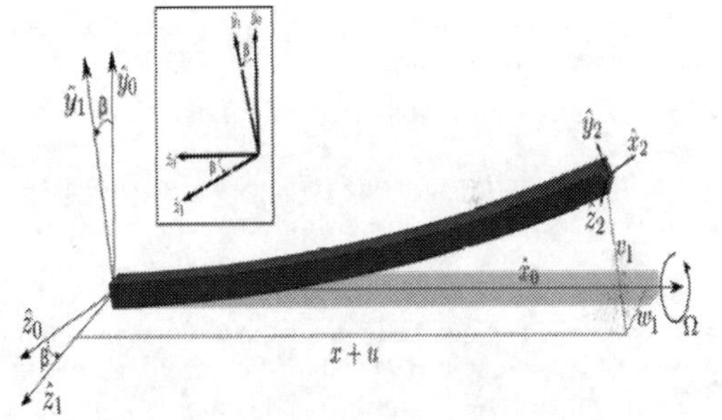

Fig. 1. A model of the spinning beam about its undeformed longitudinal direction.

Figure 1 shows the used mathematical model of the spinning beam. The axes (x_0, y_0, z_0) and (x_2, y_2, z_2) are assumed to be the principal axes of the initial undeformed and the final deformed configurations, respectively. The former axes are inertial and represent the fixed coordinate system describing the undeformed beam geometry. All the used frames have their origin on the centroid C of the given cross section of the beam. Moreover, it is assumed that the static equilibrium of the beam corresponds to the situation where the longitudinal axes x_2 and x_0 are coincident [4]. Four frames $\mathcal{F}_i$, with $i = 0, .., 3$, are necessary to describe transformations from the initial state to the final deformed state [3].

The four approaches for linearizing the equations of motion reported in [3] are

1. Approach 1: keeping all terms in the velocity vector that lead to quadratic terms in the kinetic energy.
2. Approach 2: keeping only linear terms in the velocity vector.
3. Approach 3: linearizing the Euler angles in the angular velocity vector.
4. Approach 4: linearizing the Euler angles in the velocity vector.

Approaches 1 and 4 can be seen from kinetic energy point of view as a late linearization methods and the others approaches as early linearization methods.
Equations of the transversal displacements v_1 and w_1 of the cross-section center along the rotated frame $\mathcal{F}_1$, corresponding to the four linearization approaches, can be written in the following compact non-dimensional form [3]

$$\ddot{v}_1 - \gamma_z \ddot{v}_1'' + v_1'''' - \overbrace{2r\dot{w}_1}^{Coriolis force} \overbrace{- \dot{r}(w_1 + (1-\delta_{2k})\gamma_y w_1'')}^{Euler forces} \overbrace{- r^2(v_1 + (\delta_{1k}\gamma_z - 2\delta_{3k}\gamma_y)v_1'')}^{Centrifugal forces} = 0 \quad (1)$$

$$\ddot{w}_1 - \gamma_y \ddot{w}_1'' + \Gamma w_1'''' + \overbrace{2r\dot{v}_1}^{Coriolis force} \overbrace{+ \dot{r}(v_1 - (1-\delta_{2k})\gamma_y v_1'')}^{Euler forces} \overbrace{- r^2(w_1 + (\delta_{1k}\gamma_y + 2\delta_{3k}\gamma_z)w_1'')}^{Centrifugal forces} = 0 \quad (2)$$

where δ_{ij} is the Kronecker delta, the dot and the prime are temporal and spatial derivatives, respectively. The other parameters of the model are

- k is an integer that refers to the used linearization approach.
- r is the ratio of the spinning velocity and a characteristic bending natural frequency of the beam.
- γ_y and γ_z are the inverse of the square of the slenderness about the respective axes.
- Γ is the ratio of quadratic moments with respect to y-axis and z-axis. It measures the difference in the flexural rigidities in the two principal directions of the cross section. Moreover, $\Gamma = \gamma_y/\gamma_z$ since we are studying beams with cross-sections having at least two symmetric axes.

All the four approaches equations have in common the four first terms that correspond to the linear flexural vibrations of a non-spinning Rayleigh beam model and to the Coriolis force. The differences between all the approaches are in the Centrifugal and Euler forces. Moreover, all the four approaches lead to the same Euler-Bernoulli model when rotary inertia are cancelled i.e., $\gamma_y = \gamma_z = 0$. Thus, the linearization process has no effect on the mathematical model in the case of Euler-Bernoulli beams [3].
Furthermore, the first and the third approaches have the same Euler forces terms that are caused by translational and rotational motions of the beam. The second approach has only the Euler force caused by the translational motion.
The four approaches have the same translational centrifugal forces and different contributions of the rotational motion of the beam. This latter does not contribute in the second approach when the velocity vector is taken linear.
The nonconstant spinning velocity causes the Euler forces that act as parametric excitations of the rotating beam. In the case of constant spinning velocity, approaches 2 and 4 are giving the same equations of motion.

3 Natural frequencies and stability analysis

In this section natural frequencies of the spinning Rayleigh beams are computed in the case of constant spinning velocity i.e., $\dot{r} = 0$. Thus, using separation of variables method, solutions of Eqs. (1) and (2) are sought in the following form

$$v_1(X,t) = \sum_{n=1}^{N} q_n(t)\,\phi_n(X) \quad (3)$$

$$w_1(X,t) = \sum_{n=1}^{N} p_n(t)\,\phi_n(X) \quad (4)$$

where $\phi_n(X)$ is the nth bending normal mode shape of the beam for a given boundary conditions. Then, substituting Eqs. (3) and (4) into the partial differential equations (1) and (2) leads to the following system of ordinary differential equations that governs the temporal dependencies of the solutions

$$(\phi_n - \gamma_z\,\phi_n'')\,\ddot{q}_n - 2r\phi_n\dot{p}_n + [\phi_n'''' - r^2(\phi_n + (\delta_{1k}\gamma_z - 2\delta_{3k}\gamma_y)\,\phi_n'')]\,q_n = 0 \quad (5)$$

$$(\phi_n - \gamma_y\,\phi_n'')\ddot{p}_n + 2r\phi_n\dot{q}_n + [\Gamma\phi_n'''' - r^2(\phi_n + (\delta_{1k}\gamma_y + 2\delta_{3k}\gamma_z)\,\phi_n'')]p_n = 0 \quad (6)$$

Applying the Galerkin method Eqs.(5) and (6) become

$$A_n\ddot{q}_n + B_n q_n - C_n\dot{p}_n = 0 \quad (7)$$
$$D_n\ddot{p}_n + E_n p_n + C_n\dot{q}_n = 0 \quad (8)$$

Combining Eqs.(7) and (8) we obtain the following fourth order ordinary differential equation

$$\ddddot{\alpha}_n + \eta_n\ddot{\alpha}_n + \zeta_n\alpha_n = 0 \quad (9)$$

where the variable $\alpha_n = q_n(t)$ or $p_n(t)$ and the parameters of Eq. (9) are given by

$$\eta_n = \frac{B_nD_n + C_n^2 + A_nE_n}{A_nD_n} \quad ; \quad \zeta_n = \frac{B_nE_n}{A_nD_n} \quad (10)$$

where

$$A_n = \int_0^1 (\phi_n - \gamma_z\phi_n'')\phi_n dx \quad (11)$$

$$B_n = \int_0^1 [\phi_n'''' - r^2(\phi_n + (\delta_{1k}\gamma_z - 2\delta_{3k}\gamma_y)\phi_n'')]\phi_n dx \quad (12)$$

$$C_n = \int_0^1 2r\phi_n^2 dx = 2r \quad (13)$$

$$D_n = \int_0^1 (\phi_n - \gamma_y\phi_n'')\phi_n dx \quad (14)$$

$$E_n = \int_0^1 [\Gamma\phi_n'''' - r^2(\phi_n + (\delta_{1k}\gamma_y + 2\delta_{3k}\gamma_z)\phi_n'')]\phi_n dx$$

The time dependent solutions are sought in the following form

$$\begin{aligned} q_n(t) &= a_{1n}\cos(\omega_n^+ t) + a_{2n}\sin(\omega_n^+ t) + a_{3n}\cos(\omega_n^- t) \\ &\quad + a_{4n}\sin(\omega_n^- t) \qquad (15) \\ p_n(t) &= b_{1n}\cos(\omega_n^+ t) + b_{2n}\sin(\omega_n^+ t) + b_{3n}\cos(\omega_n^- t) \\ &\quad + b_{4n}\sin(\omega_n^- t) \qquad (16) \end{aligned}$$

where the unknown amplitudes a_{in} and b_{in} (with $i = 1, .., 4$) are obtained through initial conditions.
Finally, the natural frequencies $\omega_n^\pm$ of Rayleigh spinning beams, in general framework, are given by

$$\omega_n^\pm = \frac{1}{\sqrt{2}}\sqrt{\eta_n \pm \sqrt{\eta_n^2 - 4\zeta_n}} \qquad (17)$$

The natural frequencies depend strongly on the boundary conditions through the bending mode shape $\phi_n(X)$. Furthermore, non-spinning bending frequencies of Rayleigh beams can be obtained from Eq.(17) by cancelling the spinning angular velocity i.e., $r = 0$.
Moreover, bending natural frequencies of spinning Euler-Bernoulli beams can be obtained from Eq.(17) by setting $\gamma_y = \gamma_z = 0$.
The temporal dependence $\alpha_n(t)$ of the nth mode is independent of the other modes. Thus, to perform a stability analysis, the fourth order equation (9) can be rewritten on the form of a system of four first order ODEs. The eigenvalues λ of the corresponding Jacobian matrix are given by the following characteristic equation

$$\lambda^4 + \eta_n\lambda^2 + \zeta_n = 0 \qquad (18)$$

The stability of the nth mode of the spinning beam is determined by examining the signs of real parts of the four λ solutions of Eq. (18). Hence, the divergence instability occurs for $\zeta_n = 0$ and the flutter instability occurs for $\eta_n^2 - 4\zeta_n = 0$.

4 Case study : clamped-clamped spinning beam

Approaches 1 and 2 will be applied to the spinning clamped-clamped Rayleigh beam. This case was studied by Sheu and Yang [5] and Pai et al. [6] using the approach 2 and the equations of motion were expressed in the inertial frame. The case of spinning Euler-Bernoulli beams was studied earlier by Bauer [7].
The boundary conditions corresponding to the clamped-clamped case are

$$v_1(0,t) = w_1(0,t) = v_1'(0,t) = w_1'(0,t) = 0 \qquad (19)$$
$$v_1(1,t) = w_1(1,t) = v_1'(1,t) = w_1'(1,t) = 0 \qquad (20)$$

The natural bending modes shapes of the clamped-clamped Euler-Bernoulli non spinning beam are used as the spatial dependence of the solutions. Thus,

$$\phi_n(X) = \Big[\cos(\beta_n X) - \cosh(\beta_n X) - \frac{(\cos(\beta_n) - \cosh(\beta_n))(\sin(\beta_n X) - \sin(\beta_n X))}{\sin(\beta_n) - \sinh(\beta_n)}\Big] \qquad (21)$$

where β_n is the nth positive solution of the following algebraic equation : $\cos(\beta)\cosh(\beta) = 1$.
The first four modes shapes of the 3D bending vibration of the spinning beam corresponding to Eq.(21) are shown in Fig. 2.

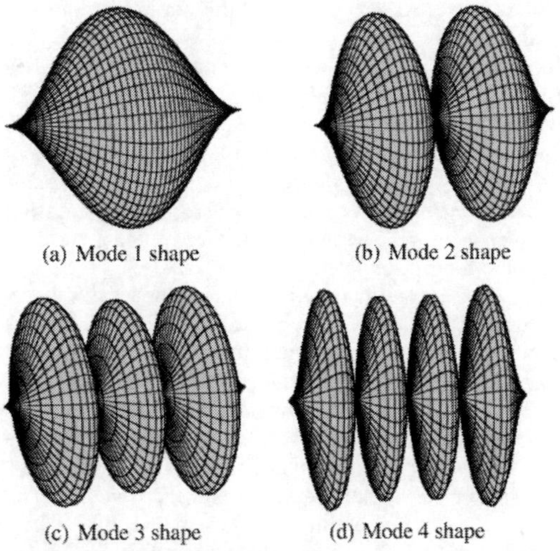

(a) Mode 1 shape (b) Mode 2 shape (c) Mode 3 shape (d) Mode 4 shape

Fig. 2. First four modes shapes of 3D bending vibration of a spinning beam.

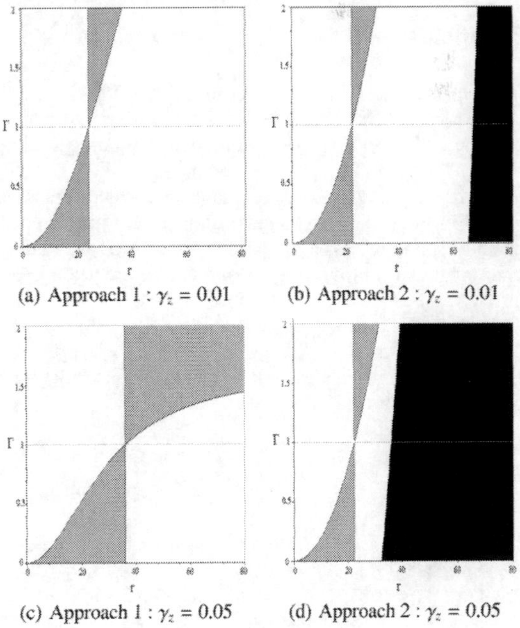

(a) Approach 1 : $\gamma_z = 0.01$ (b) Approach 2 : $\gamma_z = 0.01$ (c) Approach 1 : $\gamma_z = 0.05$ (d) Approach 2 : $\gamma_z = 0.05$

Fig. 3. Stability charts of the clamped-clamped spinning Rayleigh beam for n = 1 in the (r, Γ)-plane. The grey zone: divergence instability, the black zone: flutter instability. The left curves correspond to the approach 1 and right curves to the approach 2.

In Fig. 3 are shown the stability charts of clamped-clamped spinning Rayleigh beams, for the first mode n = 1, given by approaches 1 and 2, respectively. All these figures show that increasing γ_z i.e., decreasing the slenderness ratio parameter leads to the increase of the zones of instability (grey and black zones). Moreover, the divergence instability zone is composed of two parts that shrinks to a point at $\Gamma = 1$. All eigenvalues of this shrinking point are equal to zero.

Moreover, the approach 1 results exhibit only divergence instability, see Figs. 3a and 3c. Approach 2 exhibits both divergence and flutter instabilities, see Figs. 3b and 3d. However, its zone of divergence instability is independent of γ_z and is similar to the instability zone of the Euler-Bernoulli beam. The zone of flutter instability is increasing with increasing γ_z. The same results were found in the case of double-hinged Rayleigh beam [3].

In Fig. 4, the non-dimensional bending first natural frequencies of spinning Rayleigh and Euler-Bernoulli beams are plotted versus the non-dimensional spinning velocity r, using the linearization approaches 1 and 2, for $\gamma_z = 0.01$ and $\Gamma = 0.5$. Initially, the non spinning beams have two bending fundamental frequencies caused by the difference in bending rigidities in the y and z axes. Furthermore, ω_1^+ is the upper curve that increases with increasing r. The frequency ω_1^- is decreasing till it destabilizes by a divergence instability. Then, it stabilizes again for $r \approx 22$ and increases with increasing r. This scenario will continue for higher values of r in the case of approach 1. For the approach 2, the two fundamental frequencies of the spinning Rayleigh beam destabilize through a flutter instability. Frequencies of spinning Euler-Bernoulli beams do not exhibit a flutter instability.

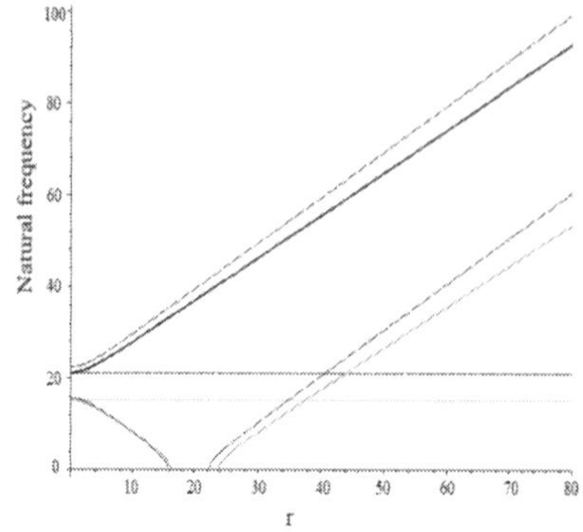

(a) Approach 1

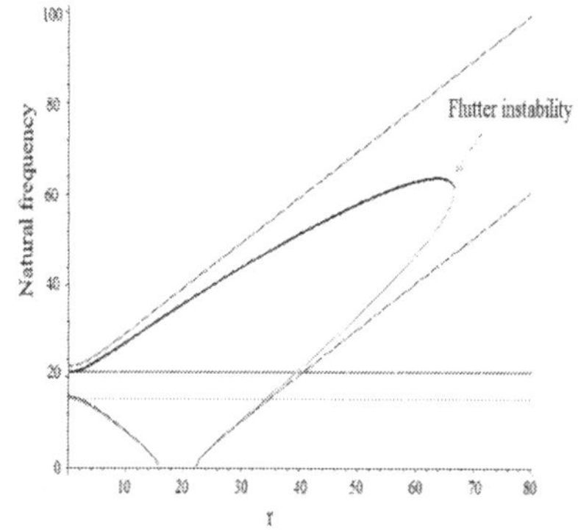

(b) Approach 2

Fig. 4. First natural frequency of clamped-clamped beam versus the spinning speed r, for $\gamma_z = 0.01$ and $\Gamma = 0.5$. The continuous lines: Rayleigh spinning beam. Dashed lines : Euler-Bernoulli spinning beam. Dotted lines: non-spinning Rayleigh beam.

5 Conclusion

In the present work the compact model developed in [3] is investigated. This model includes four linearization approaches of 3D bending linear vibrations of spinning Rayleigh beams. The natural frequencies and the instability scenarios of the free vibrations are determined in a general framework for a broad range of boundary conditions. The clamped-clamped spinning Rayleigh beam was taken as a case study. It is found that two linearization approaches are leading to different processes of destabilization of the free vibrations. The approach 1 exhibits only the divergence instability and the approach 2 exhibits divergence and flutter instabilities. Only the experimental data will be able to assess the degree of correctness of the different theoretical models.

References

1. J.S. Rao. History of rotating machinery dynamics. Springer, 2011.
2. D.F. de Castro, I.A.A. Prado, M.F.V. Pereira, D.A. Santos, J.M. Balthazar. In-Flight Parameter Estimation for Multirotor Aerial Vehicles. MATEC Web of Conferences 83, 01004, 2016
3. M.A. Aouadi, F. Lakrad. On mathematical modelling of linear flexural vibrations of spinning Rayleigh beams. J. Sound Vib. 430, 17-35, 2018.
4. A.H. Nayfeh, P.F. Pai. Linear and Nonlinear StructuralMechanics,Wiley-Interscience, New York, 2004.
5. G.J. Sheu, S.M. Yang. Dynamic analysis of a spinning Rayleigh beam, Int. J. Mech. Sci. 47, 157-169, 2005.
6. P.F. Pai, X. Qian, X. Du. Modeling and dynamic characteristics of spinning Rayleigh beams. Int. J. Mch. Sci. 68, 291-303, 2013.
7. H.F. Bauer. Vibration of rotating uniform beam. Part I: orientation in the axis of rotation, J. Sound Vib. 72, 177-189, 1980.

Effect of the localization on the response of a quasi-periodic electromagnetic oscillator array for vibration energy harvesting

Kaouthar Aouali[1,2,*], *Najib* Kacem[1], *Elyes* Mrabet[2], *Noureddine* Bouhaddi[1] and *Mohamed* Haddar[2]

[1]Univ. Bourgogne Franche-Comté, FEMTO-ST Institute, CNRS/UFC/ENSMM/UTBM, Department of Applied Mechanics, 25000 Besançon, France
[2]National Engineering School of Sfax, Research Laboratory of Mechanics Modeling and Production, Sfax, Tunisia

Abstract. Vibration energy harvesting by exploiting the multimodal approach in a quasi-periodic system is proposed. The quasi-periodic system, based on electromagnetic transduction, consists of two weakly coupled magnets mechanically guided by two elastic beams. Mistuning is achieved by varying the mechanical stiffness of one of the beams. These imperfections will lead to the vibration energy localization in regions close to the imperfections which will be exploited to maximize the harvested energy.

1 Introduction

The diversity of ambient sources of energy arouses researchers to make its scavenging a focus of interest. Many efforts have, recently, been made to develop techniques for vibration energy harvesting based on various energy conversion mechanisms. In the following, we are interested in a vibration energy harvester (VEH) by exploiting the multimodal approach [1], which involves operating multiple modes in a periodic system. Depending on the magnitude of the disorder and the strength of the internal coupling of the system, it is possible to localize the vibration energy in regions close to the imperfections [4]. Thus, we introduce an irregularity in the mechanical stiffness of one of the two beams. This phenomenon will be exploited to maximize the harvested energy. Numerical simulations have been performed to highlight the benefits of the localization on the VEH in terms of harvested power. Thus, the optimal mistuning can be determined in order to enhance the performances of the proposed device.

2 System modelling

2.1 Design

The proposed harvester, inspired from the work of Mahmoudi and al [2] and depicted in figure 1, is based on magnetic transduction. It is composed of two weakly coupled magnets guided by elastic beams. The magnetic poles are oriented in such a way that repulsive forces are created between each two adjacent magnets. These forces lead to "nonlinear magnetic stiffness". Wire-wound copper coils are wrapped around the moving magnets. When the device is subjected to a harmonic base excitation, each moving magnet oscillates around its equilibrium position and a current is induced in each coil.

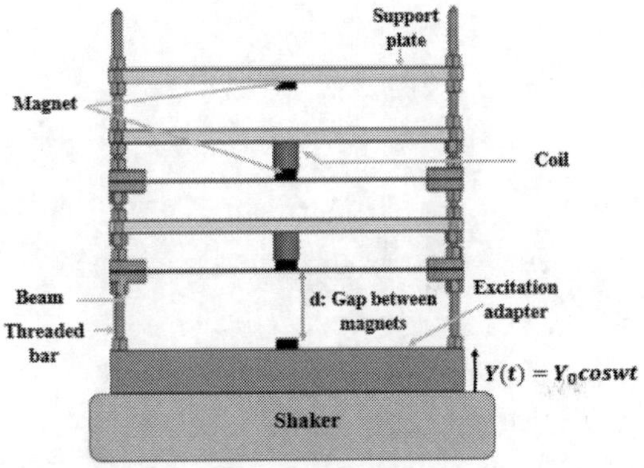

Fig. 1. Design of the proposed vibration energy harvester

The design and physical parameters are reported in Table 1.

2.2 Problem formulation

In this study, the fourth order partial differential equations of the continuum system are derived using the Hamilton principle. Then, a reduced-order model is generated by Galerkin method, transforming the system into a finite-degree-of-freedom system in terms of generalized coordinates.

3 Solution procedure

3.1. Equation of motion

* Corresponding author: aouali.kaouthar@gmail.com

After normalization and generation of a reduced order model, the coupled multi-physics problem, for a quasiperiodic system of two degrees of freedom (dof), that its equivalent model shown in figure 2, can be written in dimensionless form of modal amplitudes A_1, A_2 as:

$$\begin{cases} M_{eq1}\ddot{a}_1 + c\dot{a}_1 + (k_{mec1} + 2k_{mag})a_1 - k_{mag}a_2 = -\left(M_{eq1} + m_1\right)\ddot{Y} \\ \alpha M_{eq1}\ddot{a}_2 + c\dot{a}_2 + (k_{mec1} + 2k_{mag})a_2 - k_{mag}a_1 = -\left(\alpha M_{eq1} + m_1\right)\ddot{Y} \\ i_k(t) = \dfrac{\delta}{(R_{load} + R_{int})}a_k, \qquad k = 1,2 \end{cases} \quad (1)$$

where M_{eq1} is the equivalent mass of the kth dof, c is the sum of the mechanical and the electrical damping, k_{meci} is the mechanical stiffness, k_{mg} is the magnetic stiffness, m_1 is the mass, $\ddot{Y}$ is the base excitation, R_{load} is the load resistance and R_{int} is the internal resistance and δ is the electromechanical coupling coefficient.

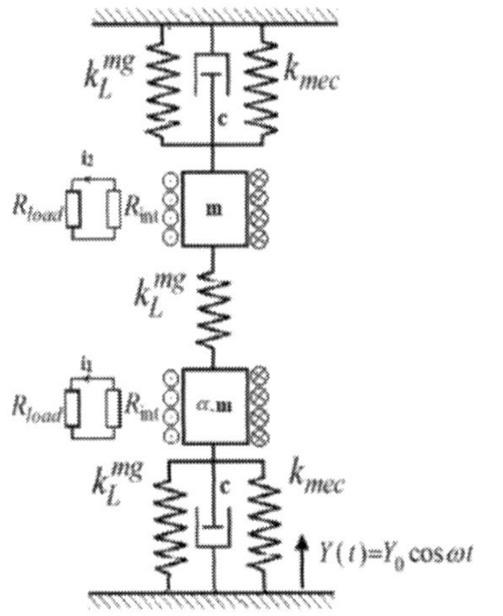

Fig. 2. Equivalent model of the proposed vibration energy harvester

The response of the system is expressed as follows:

$$\begin{cases} \begin{pmatrix} -\Omega^2 + j\Gamma\Omega + (1+2\beta) & -\beta \\ -\beta & -\alpha\Omega^2 + i\Gamma\Omega + (1+2\beta) \end{pmatrix}\begin{pmatrix} A_1 \\ A_2 \end{pmatrix} = \begin{pmatrix} (1+\mu)f\Omega^2 \\ (\alpha+\mu)f\Omega^2 \end{pmatrix} \\ i_k(t) = \dfrac{\delta}{(R_{load} + R_{int})}a_k \qquad k = 1,2\ ;\ j^2 = -1 \end{cases} \quad (2)$$

where α is the mistuning, μ is the mass ratio, β is the coupling factor, Ω is the frequency ratio, Γ is the damping factor and f is the dimensionless base excitation.

According to Mann and Sims [3], the electrical damping is defined as follows:

$$c_e = \frac{\delta^2}{R_{load} + R_{int}} \quad (3)$$

Maximum dimensionless amplitudes are expressed as follows:

$$\begin{cases} A_1 = \dfrac{(1+p)\left(-\alpha\Omega^2 + i\Gamma\Omega + 2\beta + 1\right) + \beta(\alpha + p)}{\left(-\Omega^2 + i\Gamma\Omega + 2\beta + 1\right)\times\left(-\alpha\Omega^2 + i\Gamma\Omega + 2\beta + 1\right) - \beta^2} \times Y_0\Omega^2 \\ A_2 = \dfrac{(\alpha+p)\left(-\alpha\Omega^2 + i\Gamma\Omega + 2\beta + 1\right) + \beta(1 + p)}{\left(-\Omega^2 + i\Gamma\Omega + 2\beta + 1\right)\times\left(-\alpha\Omega^2 + i\Gamma\Omega + 2\beta + 1\right) - \beta^2} \times Y_0\Omega^2 \end{cases} \quad (4)$$

3.2 Harvested power and modal localization

The induced current flowing in the load resistance provides an electric power. In order to highlight the effects of mistuning by mean of maximization of both the harvested power and the difference between maximum peaks amplitudes, we define, then, the harvested power and the modal localization ratio respectively as follows:

$$\bar{P} = R_{load}\left[\frac{\delta}{(R_{load} + R_{int})}\right]^2\left[\sum_{\alpha,\beta,R_{load}} \omega_i^2(\alpha,\beta)A_{i\max}^2(\alpha,\beta,R_{load})\right] \quad (5)$$

and

$$\tau_L = \frac{\left|A_{1\max} - A_{2\max}\right|}{max(A_{1\max}, A_{2\max})} \quad (6)$$

3 Parametric study

The variation of nondimensional amplitudes of vibrations of the two magnets with mistuning factor is depicted in figure 2. We notice that, these amplitudes are equal when $\alpha = 1$. This simulation is run with these parameters: $Y_0 = 2.\ 10^{-2}, R_{load} = 12\Omega, d = 50\ mm$.

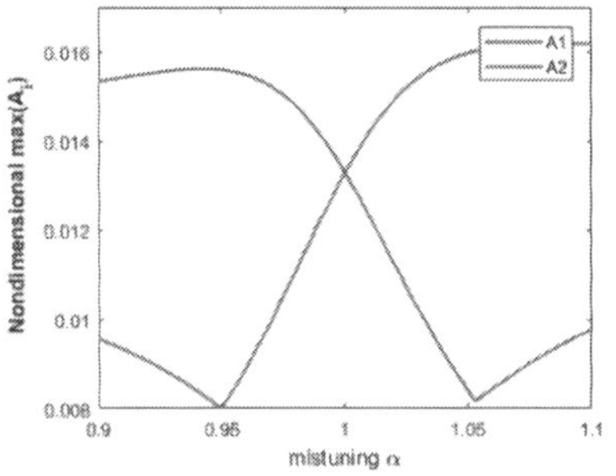

Fig. 2. Maximum dimensionless amplitudes with mistuning

In order to highlight benefits of localization, two configurations to harvest the vibration energy have been proposed. The first one aims to harvest energy from the vibrations of the two moving magnets. And, the second one aims to harvest energy from the vibrations of the mistuned magnet.

Results of the first and second configurations are shown in figures 3, 4, 5 and 6.

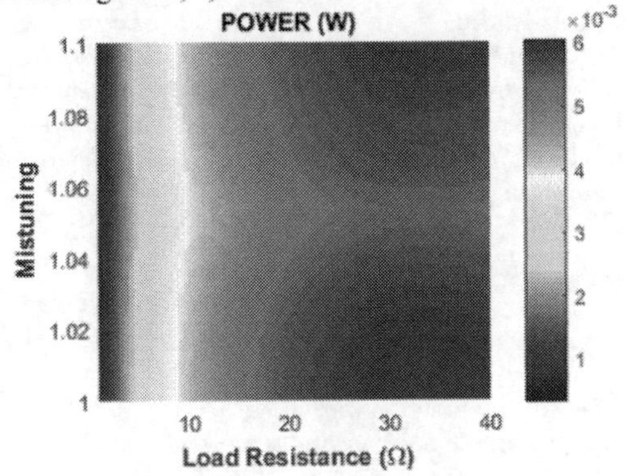

Fig. 3. Harvested power with mistuning and load resistance (configuration 1)

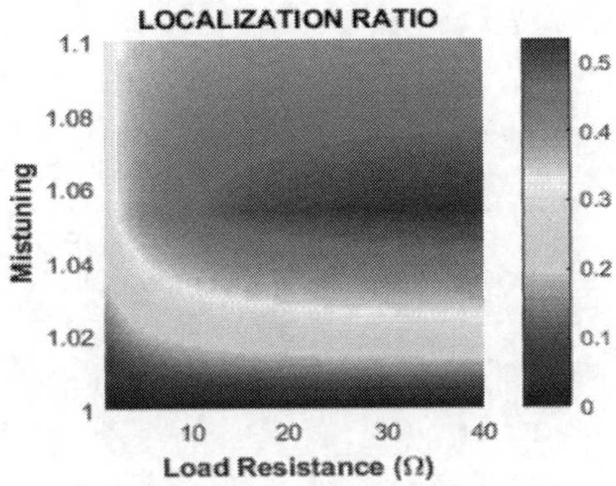

Fig. 4. Localization ratio with mistuning and load resistance (configuration 1)

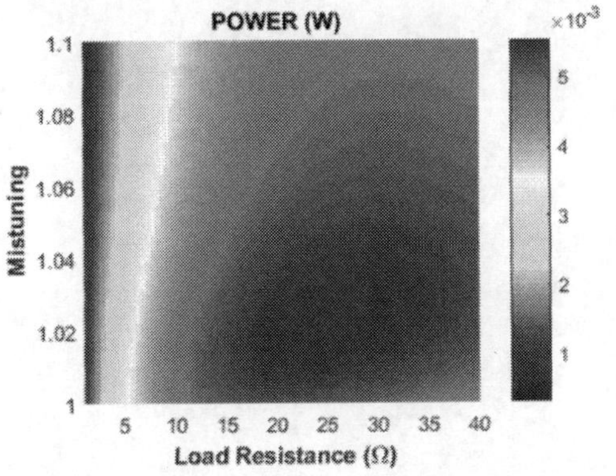

Fig. 5. Harvested power with mistuning and load resistance (configuration 2)

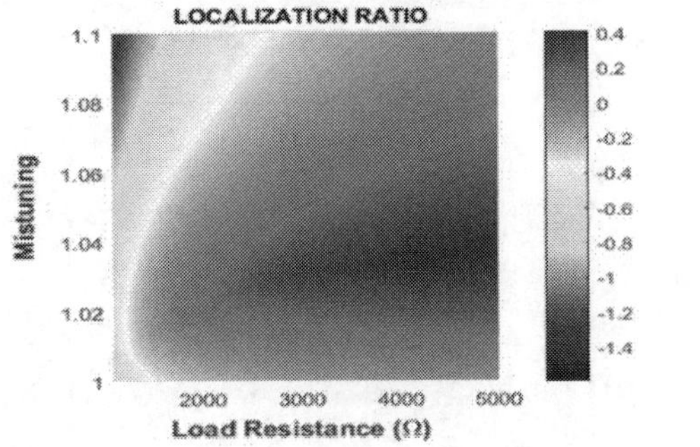

Fig. 6. Localization ratio with mistuning and load resistance (configuration 2)

For the harvested power, we notice that for the two configurations, the results are comparable. As for the modal localization, its ratio decreases slightly. The results of the second configuration compared to the first one prove the benefits of the localization phenomenon. Add to that, this configuration is more attractive in terms of technological constraints and cost. For that, it will be the configuration adopted throughout the following study.

4 Multiobjective optimization

Having already adopted the second configuration, we propose to conduct a multiobjective optimization aimed at maximizing simultaneously the harvested power and the modal localization ratio. For this purpose, we define the multiobjective problem as follows:

$$\begin{cases} \underset{\alpha,\beta,R_{load}}{Max} \ (f_1,f_2) \text{ where } \begin{cases} f_1 : \text{Harvested power} \\ f_2 : \text{Modal localization} \end{cases} \\ \alpha \le 10\% \\ 0.5\% \le \beta \le 5\% \\ R_{int} \le R_{load} \end{cases} \quad (7)$$

In the objective functions space, we plot the Pareto front which shows the variation of a function with the second one. As maximizing both functions is our purpose, we choose, then, values which ensure a compromise between the two objectives. The Pareto front is depicted in the figure 7.

* Corresponding author: aouali.kaouthar@gmail.com

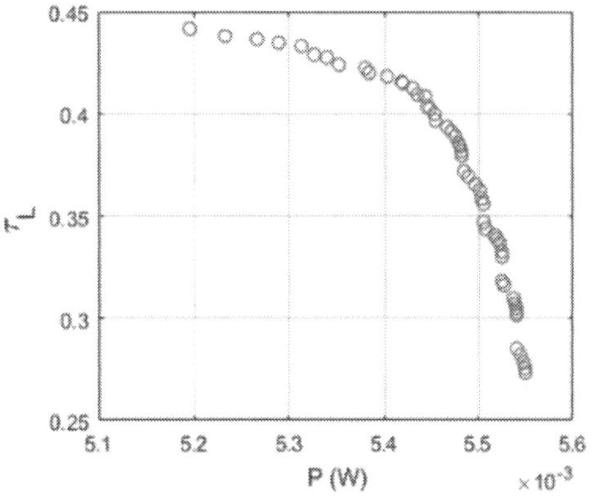

Fig. 7. The optimal solutions on the Pareto front

Then, the results of the multiobjective optimization procedure are shown in the following system:

$$\begin{cases} \underset{\alpha,\beta,R_{load}}{Max} \quad \left(\bar{P},\tau_L^{bf}\right) = (5.4mW, 41\%) \\ \alpha^* \in [1.03, 1.04] \\ \beta^* \in [1.5\%, 2.2\%] \\ R_{load}^* \in [20\Omega, 25\Omega] \end{cases} \quad (8)$$

Table 1. Design parameters

Parameter	Designation	Value
L(m)	Beam Length	$60\ 10^{-3}$
b(m)	Beam width	$4\ 10^{-3}$
Lc(m)	Magnet Length	$5\ 10^{-3}$
hs(m)	Beam thickness	$0.6\ 10^{-3}$
d(m)	Gap between magnets	$50\ 10^{-3}$
M(Kg)	Beam mass	$60\ 10^{-3}$
δ (Vs/m)	Parameter of magnet and coil	1.1
S(m)	Magnet surface	$26\ 10^{-4}$
ρ(Kg/m^3)	Steel's Density	7800
Rint (Ω)	Internal resistance	4.3
E(Pa)	Steel's Young modulus	$210\ 10^{9}$

3 Conclusions

In this paper, a vibration energy harvester based on modal localization is proposed. The concept consists in introducing a mistuning of mass in two weakly coupled magnets guided by two elastic beams. The benefits of the localization phenomenon have been investigated. A multiobjective optimization procedure was conducted in order to improve the performances of the proposed VEH with up to 5.4 mW of harvested power while maintaining a significant localization ratio of 41%.

References

1. M Y. Tadesse, S. Zhang and S. Priya, Multimodal energy harvesting system: Piezoelectric and electromagnetic. *Journal of Intelligent Material Systems and Structures,* **20**, 625-632 (2009)
2. S. Mahmoudi, N. Kacem and N. Bouhaddi, Enhancement of the performance of a hybrid non-linear vibration energy harvester based on piezoelectric and electromagnetic transductions. *Smart Materials and Structures*, **23**(7), 075024 (2014)
3. B.P. Mann, N.D. Sims, Energy harvesting from the nonlinear oscillations of magnetic levitation. *Journal of Sound and Vibration,* **319**(1-2), 515-530 (2009)
4. P.V. Malaji, S.F. Ali, Energy harvesting from near periodic structures, *Vibration Engineering and Technology of Machinery,* Mechanisms and Machine Science **23**, 411–420 (2015)

Invariant slow manifolds of an Atomic Force Microscope system under the Derjaguin-Muller-Toporov forces and a slow harmonic base motion

Sara Baghdadi, Faouzi Lakrad and Mohamed Belhaq

Laboratory of Renewable Energy and Dynamics of Systems, Faculty of Sciences Ain Chock, University Hassan II-Casablanca, Casablanca, Morocco

Abstract In the present work, we study the nonlinear vibrations of an AFM system, modeled as a linear mass-spring-damper system, under the Derjaguin-Muller-Toporov forces and subject to imposed slow harmonic base displacement. The invariant slow manifolds of the system are approximated and their bifurcations are investigated. Then, the charts of behaviors of the different operating modes of the AFM are determined. The dynamic saddle-node bifurcations of the contact and the noncontact invariant slow manifolds are found to be responsible for the occurrence of the tapping mode.

1 Introduction

Atomic Force Microscopes (AFM) are employed in a broad spectrum of applications such as imaging and characterization with an atomic resolution and in chemical and biological analysis, see for instance [1] and [2].
Atomic force microscopy is based mainly on a vibrating microcantilever with a nano-scale tip that interacts with a sample surface via short-and long-range intermolecular forces. An enhanced understanding of these vibrations is central to the correct interpretation of the AFM outputs.
Lakrad [3] and Khadraoui et al. [4] studied the nonlinear dynamics of the AFM under the Lennard-Jones force between the tip and the sample. The micro-cantilever was subject to an imposed slow harmonic displacement of its base. They computed the slow manifolds in order to explain the occurrence of the contact, noncontact and tapping modes of the AFM.
The present paper follows the same workflow than [3] but in the presence of the Derjaguin-Muller-Toporov (DMT) interaction forces between the tip and the sample. The AFM is modelled by a single degree-of-freedom lumped linear system subject to a slow harmonic base displacement. The nonlinearities are introduced in the system through tip-sample interaction forces.
The present work is organized as follows: in Section 2 the mathematical model of the AFM is presented. Section 3 contains the results of static and linear free vibration analysis. In Section 4, the theory of slow manifolds is used to determine various operating modes of the AFM.

2 Mathematical modeling

A single degree-of-freedom model, depicted in Fig.1, is used to represent a lumped-parameters model of an Atomic Force Microscope. The microcantilver is modeled as a sphere of radius R and mass m, its equivalent spring rigidity and viscous damping coefficient are denoted K and c, respectively. Moreover, the base of the microcantilver is excited by a piezoelectric actuator generating a vertical harmonic displacement $V(t) = v^* \cos(\Omega t)$.
Furthermore, the cantilever-tip-sample interaction is modeled by the van der Waals force in the non-contact phase and by the Derjaguin Muller-Toporov (DMT) forces in the contact phase. The DMT model is valid for describing stiff contacts with adhesion forces and small tip radii [5].
The equation governing the absolute displacement $x(t)$ of

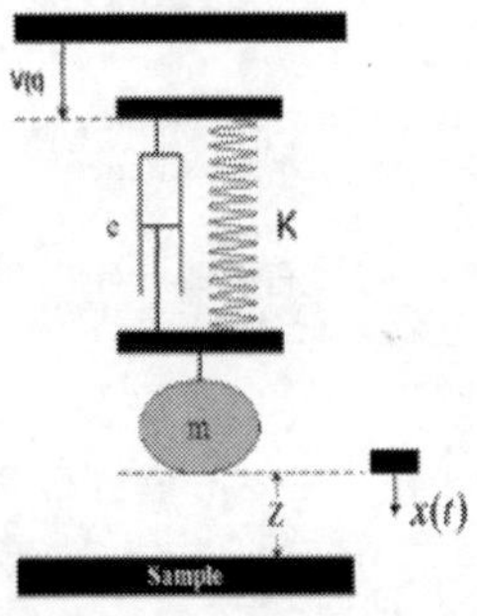

Figure 1. The studied lumped-parameters model of the AFM.

the mass towards the sample is given by

$$m\,\ddot{x}(t) + c\left(\dot{x}(t) - \dot{V}(t)\right) + K\left(x(t) - V(t)\right) = F \qquad (1)$$

where a dot indicates a derivative with respect to time t, Z is the initial gap between the AFM tip and the sample, and F is the interaction force between the mass and the sample. This latter depends piecewisely on the instantaneous gap, it is expressed as follows

$$F = \begin{cases} F_{vdW}\,, & \text{for} \quad Z - x(t) \geq a_0 \\ F_{DMT}\,, & \text{for} \quad Z - x(t) < a_0 \end{cases} \qquad (2)$$

where F_{vdW} is the non-contact attractive force of van der Waals and F_{DMT} represents the Derjaguin-Muller-Toporov contact force. The DMT force is composed of the long range attractive van der Waals and the short-range repulsive Hertzian contact forces [6]. These forces are given by

$$F_{vdW} = \frac{HR}{6(Z - x(t))^2} \quad (3)$$

$$F_{DMT} = \frac{HR}{6a_0^2} - \frac{4}{3}E^* \sqrt{R}[a_0 - (Z - x(t))]^{\frac{3}{2}} \quad (4)$$

where a_0 represents the intermolecular distance at which the contact is initiated; H, R, and E^* are, respectively, the Hamaker contant, the tip radius, and the effective modulus of elasticity. The latter is given by

$$\frac{1}{E^*} = \frac{1 - \nu_m^2}{E_m} + \frac{1 - \nu_s^2}{E_s} \quad (5)$$

with ν_m and E_m are, respectively, the Poisson ratio and modulus of elasticity of the micro-cantilever and similarly ν_s and E_s are those of the sample.
The equation of motion (1), written in nondimensional form, can be expressed as follows

$$y''(\tau) + 2\xi y'(\tau) + y(\tau) = (1 - \frac{V(\tau)}{Z}) - 2\xi\frac{V'(\tau)}{Z} - \tilde{F} \quad (6)$$

where distances were normalized by the initial gap Z.

- τ is a new scale of time : $\tau = \omega\, t$, with $\omega = \sqrt{K/m}$ is the linear natural frequency of the micro-cantilever in the absence of the sample interaction forces. The prime is the derivative with respect to τ.
- ξ is the damping ratio: $\xi = c/(2m\omega)$.
- y is the nondimensional instantaneous tip/sample gap: $y = (Z - x)/Z$.
- The nondimensional imposed displacement $V(\tau)/Z = f\cos(\epsilon\tau)$, where $f = v^*/Z$ and $\epsilon = \Omega/\omega$.

The tip/sample interaction force can be rewritten

$$\tilde{F} = \begin{cases} \frac{\gamma_2}{y(\tau)^2} & \text{for} \quad y(\tau) \geq A_0 \\ \frac{\gamma_2}{A_0^2} - \gamma_1[A_0 - y(\tau)]^{\frac{3}{2}} & \text{for} \quad y(\tau) < A_0 \end{cases} \quad (7)$$

$$\text{where } \gamma_1 = \frac{4}{3}\frac{E^*\sqrt{RZ}}{K} \;;\; \gamma_2 = \frac{HR}{6KZ^3} \text{ and } A_0 = \frac{a_0}{Z}.$$

3 Static equilibria and free vibrations

In the present section we will compte the static equilibria of Eq.(6) and the natural frequencies of the free vibrations around them.

3.1 Static equilibria

The static study of the model is obtained by dropping the time derivatives and cancelling the base motion in Eq.(6). Then, the static equilibria y_s are obtained by solving the following third order algebraic equations

$$\begin{cases} y_s^3 - y_s^2 + \gamma_2 = 0 & \text{for} \quad y_s \geq A_0 \\ y_s = 1 - \frac{\gamma_2}{A_0^2} + \gamma_1(A_0 - y_s)^{\frac{3}{2}} & \text{for} \quad y_s < A_0 \end{cases} \quad (8)$$

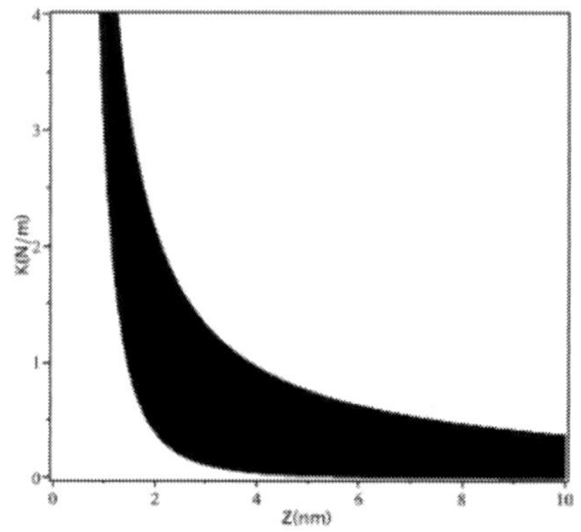

Figure 2. Number of static equilibria of Eqs. (8) in the (Z, K)-plane. The black zone corresponds to the existence of three equilibria, and the white zone to one equilibrium.

Figure 2 shows the number of static equilibria in the plane of the stiffness K of the system versus the initial gap Z. The black zone corresponds to the existence of three static equilibria and the white zone to the existence of one static equilibrium. In fact, the contact zone i.e., $y_s < A_0$ has only one static equilibrium, while the non-contact zone $y_s \geq A_0$ can have three solutions for $\gamma_2 < \frac{4}{27}$.
It is worth noting that the system can have one static equilibrium for hard micro-cantilevers and for high initial gaps. Figure 3 shows, for $K = 0.89$ N/m, the static mass-sample

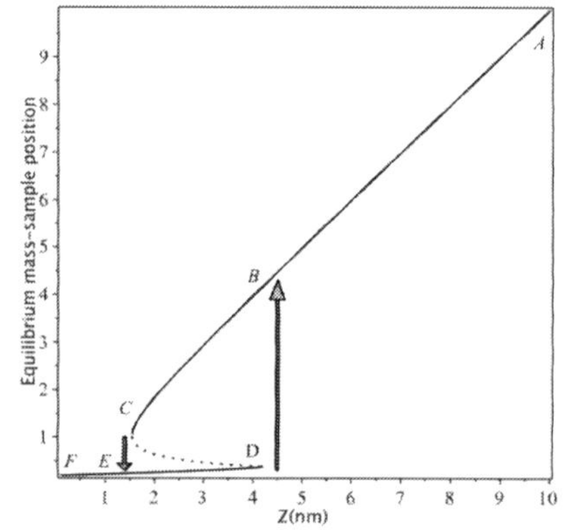

Figure 3. Static tip/sample separation versus Z.

separation distance versus the initial gap Z. The resulting static curve can be divided into two regions; a bistable region, for Z ∈ [1.5nm, 4.2nm], formed by the stable branches BC, ED and the unstable branch CD and a monostable region formed by the stable branches AB and EF.
The bistable region is delimited by two saddle-node bifurcations at points C and D where jumps occur when varying the initial gap Z.

3.2 Free vibrations

In order to compute the natural frequencies of the system under the effects of the interaction forces, the nondimen-

sional Eqs.(6) and (7) are linearized around a given stable static equilibrium y_s. Moreover, the damping and the imposed base motion are cancelled.
The normalized natural frequencies of the contact mode $freq_c$ and the noncontact mode $freq_{nc}$ are given by

$$freq_c = \frac{1}{2\pi}\sqrt{1 + \frac{3\gamma_1}{2}\sqrt{A_0 - y_s}} \quad \text{for} \quad y_s < A_0 \quad (9)$$

$$freq_{nc} = \frac{1}{2\pi}\sqrt{1 - 2\frac{\gamma_2}{y_s^3}} \quad \text{for} \quad y_s \geq A_0 \quad (10)$$

In the absence of the interaction forces, the contact and noncontact frequencies are constant and equal to $1/(2\pi)$. In Fig. 4 are shown, versus the initial gap Z, numerical and

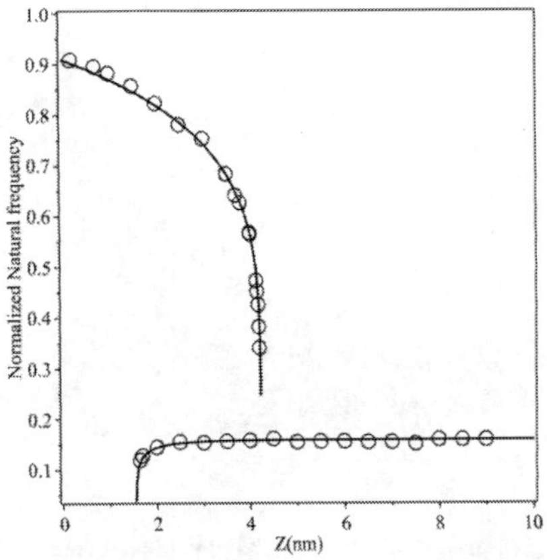

Figure 4. Normalized natural frequency versus the initial gap Z. Circles are obtained numerically from Eq.(6). Continuous lines are analytically computed frequencies in Eqs. (9) and (10).

analytical normalized natural frequencies given by equations (9) and (10), respectively, for contact and noncontact modes. The numerically computed frequencies are obtained by integrating Eq.(6) using a fourth order Runge-Kutta algorithm, then the corresponding power spectrum is computed using a Fast Fourier Transformation algorithm.
Figure 4 shows that in the contact phase, the natural frequency (9) is decreasing with increasing the initial gap Z. This can be explained by the decrease of the apparent stiffness of the micro-cantilever. In the non-contact phase, increasing the initial gap Z leads to the increase of the natural frequency (10) till reaching the natural frequency of the system without any force of interaction i.e., $1/2\pi$. In short we have two zones: one with one solution corresponding to monostable region, and the other zone with two stable solution that represents the bistable region.

4 Dynamic behavior

In the present paper, the frequency of the base displacement is taken very small with respect to the natural frequency of the system i.e., $\epsilon = \Omega/\omega << 1$. The case of resonant excitations was investigated in many papers, see for instance [7].
Equation (6) can be written as a fast-slow system since it is ruled by two dynamics: the first on the time scale of order 1 (the natural frequency) and the second one on the time scale of order ϵ (the base displacement).
Introducing the new slow time scale $T = \epsilon\tau$, Eq.(6) becomes

$$\epsilon^2\frac{d^2y}{dT} + 2\epsilon\xi\frac{dy}{dT} + y = 1 - f\cos(T) + 2\epsilon\xi f\sin(T) - \tilde{F}(T) \quad (11)$$

In the limit $\epsilon \to 0$ one can compute an approximation of the slow manifolds by solving the following algebraic equations

$$y^3 + (f\cos(T) - 1)\,y^2 + \gamma_2^2 = 0, \quad \text{for} \quad y \geq A_0 \quad (12)$$

$$y - \gamma_1(A_0 - y)^{\frac{3}{2}} + f\cos(T) = 1 - \frac{\gamma_2}{A_0^2}, \quad \text{for } y_s < A_0 (13)$$

The invariant slow manifolds can be viewed as periodic quasi-static lines of equilibria with respect to the slow time scale T. During a period of T they can bifurcate giving rise to very interesting phenomena [8]. Indeed, the computation of the invariant slow manifolds and their bifurcations are of great importance for determining and understanding the dynamics of the AFM system ruled by Eq.(1). In fact, solutions of the system will spend most of their time following attracting slow manifolds. This phenomenon is sometimes called "slaving principle" or adiabatic reduction [9].
Figure 5 shows the graphs of the slow manifolds versus the slow time T, for various values of the amplitude f of the imposed harmonic displacement of the base and for an initial gap $Z = 3$ nm, belonging to the bistable static region, see Fig. 4. The stable manifolds are plotted in continuous lines and unstable ones with dashed lines. The lower and the upper stable solutions correspond to contact and noncontact slow manifolds, respectively.
In Fig. 5a, for $f = 0$ i.e., no base motion, the slow manifolds are the coexisting three static equilibria, that are independent of T, corresponding to the contact and noncontact modes and separated by an unstable solution. When increasing f the slow manifolds are becoming periodic with increasing amplitudes. Thus, for $f = 0.39$ the contact slow manifold collides with the unstable manifold leading to a sadlle-node bifurcation. Consequently, for $f = 0.45$ the contact slow manifold does not exist for all times and the noncontact slow manifold is the visited solution. Increasing further f leads to a new collision between the noncontact slow manifold and the unstable slow manifold. Hence, for $f = 0.55$, the contact and the non contact slow manifolds will be both visited during a period of the base displacement. This solution is geometrically a periodic burster and physically corresponds to a tapping mode of the AFM [3]. The confirmation that the solutions of the nondimensional Eq.(6) are following the scenario discussed previously is shown in Fig.6 for $Z = 3$ nm.
Based on the computed slow manifolds, Fig. 7 shows the behavior chart of the system in the plane (f, Z). In the white region there exist only one solution: either the contact solution in the lower zone or the nonconact solution in the upper zone. In the light grey zone, the two solutions coexist, but only one is chosen depending on the initial conditions. In the dark grey zone, the two solutions will be visited during a period of the base motion: the tapping mode.

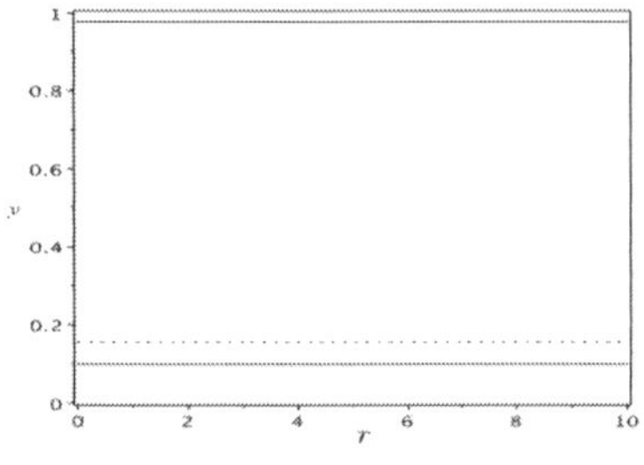

(a) $f = 0$

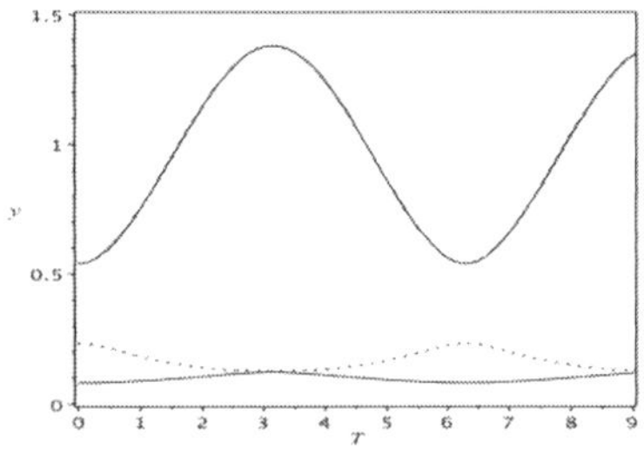

(b) $f = 0.39$

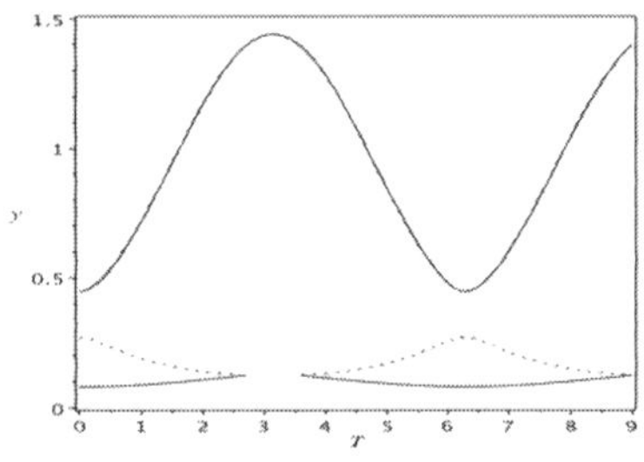

(c) $f = 0.45$

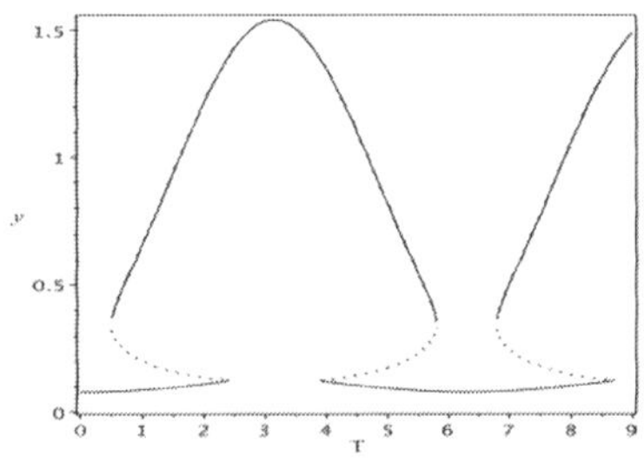

(d) $f = 0.55$

Figure 5. Graphs of the slow manifolds, given by Eqs.(12) and (13), for various base amplitudes f and for Z=3 nm

5 Conclusion

In the present work, we studied nonlinear dynamics of a lumped single d.o.f. model of an AFM system under the vand der waals and Derjaguin-Muller-Toporov interaction forces with very low frequency displacement of its base. The invariant slow manifolds theory is used to determine the different operating modes of the AFM (contact, non-contact and tapping modes). It is shown that the tapping mode is triggered by dynamic saddle-node bifurcations.
As a continuation of the present work, the nonlinear dynamic of continuous models of the AFM will be studied under the effect of imposed slow harmonic displacements.

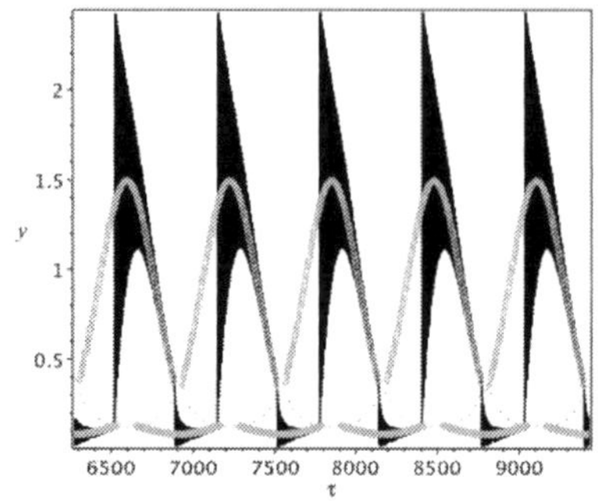

Figure 6. Numerical solutions of Eq.(6) in black and slow manifolds in grey, for f = 0.5 and Z = 3 nm.

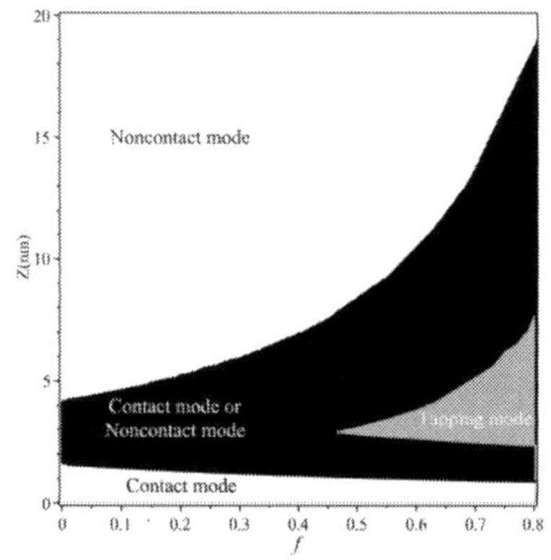

Figure 7. Chart of the system behaviors in the plane (f, Z).

References

1. G. Binnig, C.F. Quate, C. Gerber. Atomic force microscope. Phys Rev Lett 56(9): 930-4, 1986.
2. P. Eaton, P. West. Atomic force microscopy. Oxford University Press, 2010.
3. F. Lakrad. Invariant slow manifolds of an Atomic Force and a slow harmonic base motion. Commun Nonlinear Sci Numer Simulat 32, 49-62, 2016.
4. M. Khadraoui , F. Lakrad, M. Belhaq. Effecrs of a slow harmonic displacement on an Atomic Force Microscope system under Lennard-Jones forces, MATEC Web of conferences 83, 04001, 2016.
5. R. Garcia. Amplitude modulation atomic force microscopy. Wiley-VCH Verlag, 2010.
6. J. Israelachvili. Intermolecular and surface forces. 2nd ed. Academic; 1991.
7. I. Manoubi, F. Najar, S. Choura, A.H. Nayfeh. Nonlinear Dynamical analysis of an AFM tapping mode microcantilever beam. MATEC Web of Conferences 1, 04002, 2012.
8. F. Lakrad, W. Schiehlen. Effects of a low frequency parametric excitation. Chaos Solitons & Fractals 22: 1149-64, 2004.
9. N. Berglund, B. Gentz. Noise-induced phenomena in slow-fast dynamical systems. London: Springer-Verlag; 2006.

Natural Convection in a Horizontal Porous Annulus under Quasi-periodic Gravitational Modulation

Jabrane Belabid[1,a], Karam Allali[1,b], and Mohamed Belhaq[2,c]

[1] University Hassan II-Casablanca, FSTM, Department of Mathematics, PO Box Mohammadia, Morocco
[2] University Hassan II-Casablanca, FS, Department of Physics, Casablanca, Morocco

Abstract. The effect of quasi-periodic gravitational modulation on the natural convection in a horizontal porous annulus is studied in this paper. The model we consider includes the heat and the hydrodynamics equations under Darcy law. The derived model with the stream function is solved numerically using the alternate direction implicit method. Results show that the convective instability and the flow regime are influenced by frequencies ratio of the quasi-periodic gravitational modulation.

1 Introduction

Natural convection and heat transfer in porous media has been motivated by a wide range of engineering application such as cooling of electronic equipments, aircraft cabin insulation, geothermal systems, underground transmission lines, storage of nuclear waste materials, solidification of casting to name just few (see for instance [1–3], and references therein).

The pioneer work of Caltagirone [4] studied the free convection flows in a porous horizontal annulus heated from the inner cylinder. The author observed different flow regimes when the Rayleigh number exceeds some critical value. Later, many authors tackled the same problem of natural convection in porous annulus. For example, Rao et al. [5] studied the phenomena of two-dimensional bifurcation using the Galerkin method to solve the governing equations. The critical Rayleigh numbers determined numerically are in a very good agreement with the experimental results and with those of Caltagirone [4]. Himasekhar and Bau [6] investigated the problem using the regular perturbation expansion for different radii ratio.

Recently Belabid and Allali [7] studied the effect of periodic gravitational modulation on convective instability in horizontal annulus. It was shown that the convective instability depends on the modulation amplitude and frequency. Although, the influence of periodic gravitational modulation on convective instability is widely studied, only few works considered the effect of quasi-periodic (QP) gravitational modulation on the convective instability. For instance, it was shown than the frequencies ratio has a stabilizing or destabilizing effect on the onset of convection for the problems dealing with reaction front propagation [8,9] and for the problems of Hele-shaw cell [10,11]. Motivated by the recent work by Belabid and Allali [7], the main objective of the present paper is to examine the effect of the QP gravitational modulation on the convective instability in a porous horizontal annulus. To this end, we assume that the annulus is subject to QP gravitational modulation with two incommensurate frequencies.

The rest of the work is organized as follows. The problem description is given in the next section. The numerical method is illustrated in Section 3, and in Section 4 numerical results are given. Section 5 concludes the work.

2 Problem description and mathematical formulation

We consider a porous layer bounded by two horizontal concentric cylinders saturated with a Newtonian fluid as shown in Fig. 1. Both inner and outer cylinders are kept at uniform, constant and different temperature T_i and T_o respectively with $T_i > T_o$. It is taken into account that the

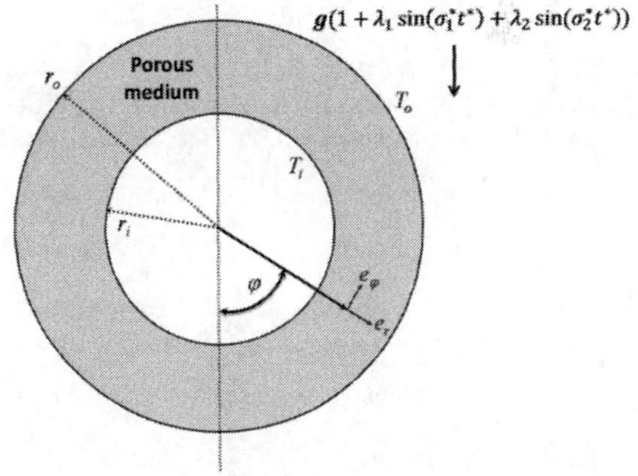

Fig. 1. Sketch of the problem

flow is two-dimensional, steady and laminar. The porous medium is considered to be homogenous, isotropic and subjected to a QP gravitational modulation with two incommensurate frequencies. Under the Boussinesq approximation, the basic steady conservation of thermal energy,

[a] e-mail: belabide@gmail.com
[b] e-mail: allali@hotmail.com
[c] e-mail: belhaq@yahoo.fr

momentum and mass equations in polar coordinates system can be written as:

$$\frac{\partial T^*}{\partial t^*} + (\mathbf{V}^*.\nabla)T^* = \alpha\nabla T^* \quad (1)$$

$$\mathbf{V}^* = -\frac{K}{\mu}(\nabla P^* - \rho\mathbf{g}(1 + \lambda_1 \sin(\sigma_1^* t^*) + \lambda_2 \sin(\sigma_2^* t^*))) \quad (2)$$

$$\nabla.\mathbf{V}^* = 0 \quad (3)$$

with the following boundary conditions:
On the inner cylinder surface: $u^* = 0,\ T^* = T_i$.
On the outer cylinder surface: $v^* = 0,\ T^* = T_o$,
where T^* is the temperature, V^* is the velocity, P^* is the pressure, α is the diffusion, K the permeability, μ the viscosity, ρ the density, λ_i and σ_i^* stand for the amplitudes and the frequencies of the QP modulation, respectively (i=1,2).

To obtain the dimensionless model, we introduce the following transformation formula:
$r = \frac{r^*}{r_i}, \mathbf{V} = \frac{\mathbf{V}^*}{\alpha}r_i, T = \frac{T^* - T_o}{T_i - T_o}, t = \frac{t^*}{r_i^2}\alpha, P = \frac{P^* K}{\alpha\mu}$,
$\sigma_1 = \frac{\sigma_1^* r_i^2}{\alpha}$ and $\sigma_2 = \frac{\sigma_2^* r_i^2}{\alpha}$. Therefore, the dimensionless governing equations can be rewritten as:

$$\frac{\partial T}{\partial t} + (\mathbf{V}.\nabla)T = \nabla T \quad (4)$$

$$\mathbf{V} = -\nabla P - RaT\boldsymbol{k}(1 + \lambda_1 \sin(\sigma_1 t) + \lambda_2 \sin(\sigma_2 t)) \quad (5)$$

$$\nabla.\mathbf{V} = 0 \quad (6)$$

where Ra is the Rayleigh number given by $Ra = \frac{g\beta K\Delta T r_i}{\alpha\nu}$; and $\sigma_2 = \eta\sigma_1$. The dimensionless boundary conditions are as follows:
On the inner cylinder surface: $u = 0, T = 1$.
On the outer cylinder surface: $v = 0, T = 0$.
Due to the physical symmetry of the problem the following boundary conditions are added:
$\varphi = 0, \pi : v = 0, \frac{\partial T}{\partial \varphi} = 0$
The governing equations for the laminar vibrated natural convection in terms of stream function-temperature formulation can be expressed as follows:

$$\nabla^2\psi = -Ra(\sin\varphi\frac{\partial T}{\partial r} + \frac{\cos\varphi}{r}\frac{\partial T}{\partial \varphi})(1+\lambda_1 \sin(\sigma_1 t)+\lambda_2 \sin(\sigma_2 t)) \quad (7)$$

$$\frac{\partial T}{\partial t} + \frac{1}{r}\frac{\partial \psi}{\partial \varphi}\frac{\partial T}{\partial r} - \frac{1}{r}\frac{\partial \psi}{\partial r}\frac{\partial T}{\partial \varphi} = \nabla^2 T \quad (8)$$

and the corresponding boundary conditions take the following form:
$r = 1 : T = 1$ and $\frac{\partial \psi}{\partial \varphi} = 0$;
$r = R : T = 0$ and $\frac{\partial \psi}{\partial \varphi} = 0$;
$\varphi = 0, \pi : \frac{\partial T}{\partial \varphi} = \frac{\partial \psi}{\partial r} = 0$.

The local Nusselt number along the inner and outer cylinders are estimated as the ratio of convective to conductive heat transfer:

$$Nu_i(r = 1, \varphi) = -\ln R\left.\frac{\partial T}{\partial r}\right|_{r=1} \quad (9)$$

$$Nu_o(r = R, \varphi) = -R\ln R\left.\frac{\partial T}{\partial r}\right|_{r=R} \quad (10)$$

The average Nusselt numbers evaluated at the inner and outer cylinders can be written as:

$$\overline{Nu} = \frac{1}{\pi}\int_0^\pi Nu_i\,d\varphi = -\frac{1}{\pi}\ln R\int_0^\pi \left.\frac{\partial T}{\partial r}\right|_{r=1} d\varphi \quad (11)$$

$$\overline{Nu} = \frac{1}{\pi}\int_0^\pi Nu_o\,d\varphi = -\frac{1}{\pi}R\ln R\int_0^\pi \left.\frac{\partial T}{\partial r}\right|_{r=R} d\varphi \quad (12)$$

For a sufficient refined mesh, both expressions of the average Nusselt number converge to the same value.

3 Numerical methods and code validation

The alternate direction implicit finite difference method is used to solve the mathematical problem along with boundary conditions. The Thomas algorithm in conjunction with iterations is used to solve the resulting matrix systems. The convergence of our in-house numerical code has been assessed by a mesh testing procedure. In fact, numerical tests, using various mesh grids, were examined for the case without any modulation i.e. $\lambda_i = 0,\ i = 1, 2$; in order to determine the best compromise between the grid independent solutions and the calculation time. The iteration process is terminated when the following criterion is satisfied in each node of the grid:

$$\max\left|\frac{\xi_{i,j}^{n+1} - \xi_{i,j}^n}{\xi_{i,j}^n}\right| < 10^{-8}$$

	Grid size	$Ra = 50$	$Ra = 100$
Caltagirone [4]	49×49	1.328	1.829
Facas and Farouk [14]	25× 25	1.362	1.902
Rao et al. [6]	10× 10	1.341	1.861
Facas [15]	50× 50	1.342	1.835
Charrier-Mojtabi [16]	30× 95	1.344	1.867
Mota et al. [17]	161× 101	1.338	1.861
Alfahaid et al. [12]	10× 18	1.317	1.865
Alloui and Vasseur [13]	100× 240	1.343	1.868
This study	49× 49	1.343	1.851

Table 1. Nusselt number for $R = 2$. Comparison with literature

where ξ refers to T or ψ, the subscripts i and j indices denote grid location in the (r, φ) plane and n designates the iteration number. The present code was validated by comparing the obtained result for $Ra = 50$ and $Ra = 100$ with the already published data (see Table 1). Figure 2 shows that we can choose the mesh grid 71×71 for our numerical simulations which ensures the mesh grid independence.

4 Results and discussions

The numerical simulation will be performed for the fixed amplitude $\lambda_1 = \lambda_2 = 0.1$, the frequency $\sigma_1 = 10$ and the radius $R = 2$. The frequencies ratio η and the Rayleigh number will be varied adequately in order to check the QP

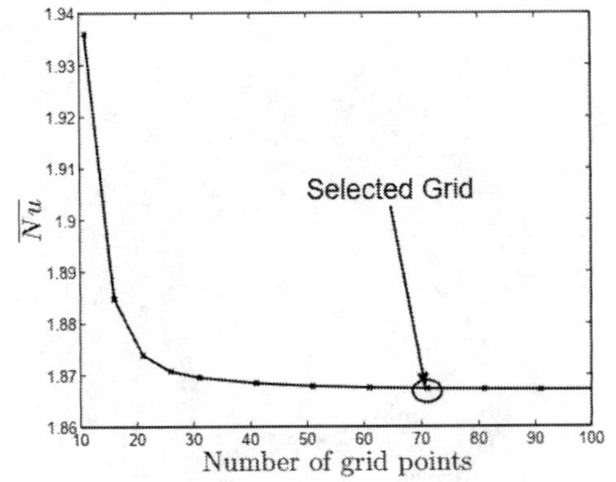

Fig. 2. Mesh effect on $\overline{Nu}$ for $\lambda_1 = \lambda_2 = 0$ and $Ra = 100$

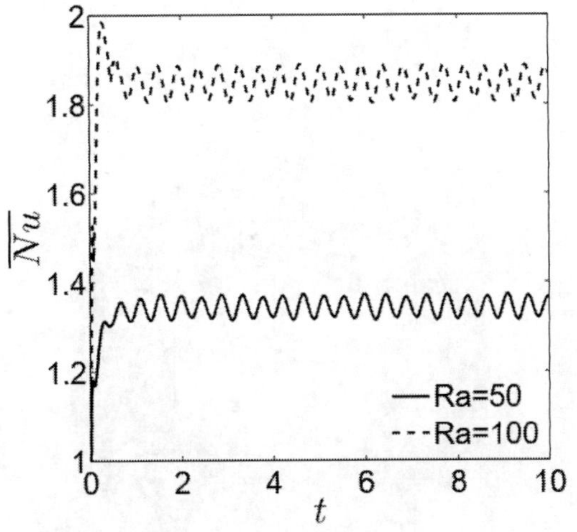

Fig. 3. Nusselt number as function of time for $\eta = \sqrt{2}$

modulation on the onset of convection. Figure 3 shows the Nusselt number as function of time for different Rayleigh number. First, we clearly observe a small modulation in the behaviour of the Nusselt number. On the other hand, increasing Rayleigh number leads to an increase of the heat transfer. The comparison between the case of periodic and QP gravitational modulation is illustrated in Fig. 4. Indeed, when the two frequencies ratio is zero we find exactly the results in [7]. By increasing the value of η the heat transfer has an alternate behaviour around a mean value which means that a destabilizing effect of the QP gravitational modulation is observed. Finally, in Fig. 5, we observe a stabilizing effect for small frequencies ratio and a destabilizing effect for high ones since transition from unicellular to bicellular flow regime occurs by increasing the frequencies ratio.

5 Conclusion

In this paper, we have studied the effect of the QP gravitational modulation on convective instability in horizontal porous annulus. The gravitational modulation is assumed

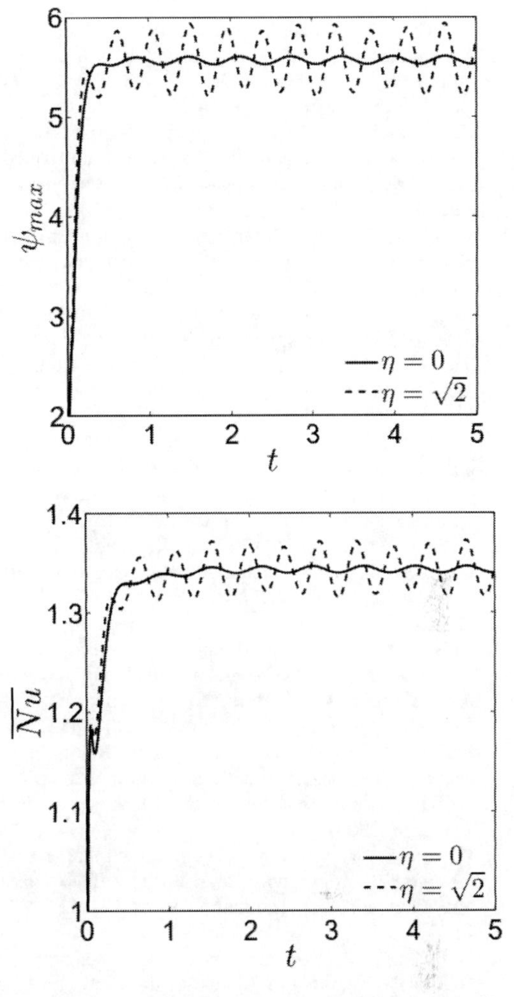

Fig. 4. Nusselt number and ψ_{max} as function of time for $Ra = 50$

to be with two incommensurate frequencies. The considered model consists of heat equation coupled with the hydrodynamics under Darcy law. The reduced model with the stream function formulation was solved using the alternate direction implicite method. It was shown that the convective instability depends on the frequencies ratio. More precisely, it was observed that the QP gravitational modulation has a destabilizing effect induced by an alternate behaviour of the Nusselt number.

References

1. P. Vadasz, Emerging topics in heat and mass transfer in porous media, Springer-Verlag, New York, 2008.
2. D.A. Nield, A. Bejan, Convection in porous media, 4th ed., Springer-Verlag, New York, 2013.
3. K. Vafai, Handbook of porous media, 3rd ed., Taylor & Francis, New York 2015.
4. J.P. Caltagirone, Instabilities in a porous medium bounded by two concentric horizontal cylinders, Journal of Fluid Mechanics **76** (1976) 337–362.

 https://doi.org/10.1051/matecconf/201824101005

5. Y.F. Rao, K. Fukuda, S. Hasegawa, Steady and transient analysis of natural convection in a horizontal porous annulus with the Galerkin method, Journal of Heat Transfer **109** (1987) 919–927.
6. K. Himasekhar, H.H. Bau, Two-dimensional bifurcation phenomena in thermal convection in horizontal, concentric annuli containing saturated porous media, Journal of Fluid Mechanics **187** (1988) 267–300.
7. J. Belabid, K. Allali, Influence of gravitational modulation on natural convection in a horizontal porous annulus, Journal of Heat Transfer **139** (2017) 022502.
8. K. Allali, M. Belhaq, K. El Karouni, Influence of quasi-periodic gravitational modulation on convective instability of reaction fronts in porous media, Communications in Nonlinear Science and Numerical Simulation, **17** (2012) 1588–1596.
9. K. Allali, S. Assiyad, M. Belhaq, Convection of polymerization front with solid product under quasi-periodic gravitational modulation, Nonlinear Dynamics and Systems Theory, **14** (2014) 323–334.
10. T. Boulal, S. Aniss, M. Belhaq, R. Rand, Effect of quasiperiodic gravitational modulation on the stability of a heated fluid layer, Physical Review E, **76** (2007) 056320.
11. T. Boulal, S. Aniss, M. Belhaq, A. Azouani, Effect of quasi-periodic gravitational modulation on the convective instability in Hele-Shaw cell, International Journal of Non-Linear Mechanics **43** (2008) 852–857.
12. A.F. Alfahaid, R.Y. Sakr, M.I. Ahmed, Natural convection heat transfer in concentric horizontal annuli containing saturated porous media, IIUM Engineering Journal **6** (2005) 41–54.
13. Z. Alloui, P. Vasseur, Natural convection in a horizontal annular porous cavity saturated by a binary mixture, Computational Thermal Sciences **3** (2011) 407–417.
14. G.N. Facas, B. Farouk, Transient and Steady State Natural Convection in a Porous Medium between Two Concentric Cylinders, Journal of Heat Transfer **105** (1983) 660–663.
15. G.N. Facas, Natural Convection From a Buried Pipe with External Baffles, Numerical Heat Transfer, Part A **27** (1995) 595–609.
16. M.C. Charrier-Mojtabi, Numerical simulation of two-three dimensional free convection flows in a horizontal porous annulus using a pressure and temperature formulation, International Journal of Heat and Mass Transfer **40** (1997) 1521–1533.
17. J.P.B. Mota, I.A.A.C. Esteves, C.A.M. Portugal, J.M.S.S. Esperana, E. Saatdjian, Natural convection heat transfer in horizontal eccentric elliptic annuli containing saturated porous media, International Journal of Heat and Mass Transfer **36** (2000) 4367–4379.

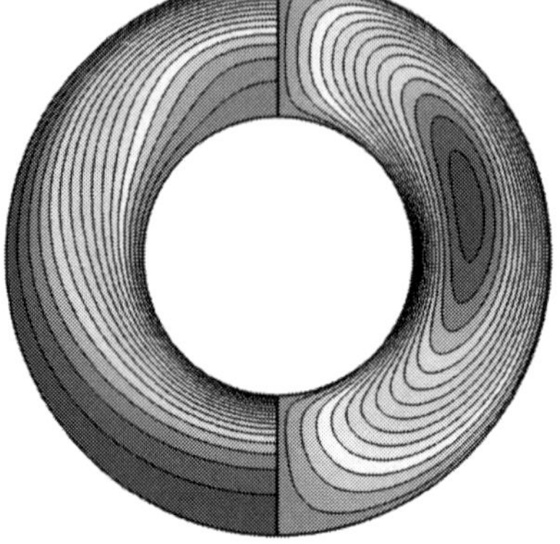

(a): $\eta = \sqrt{2}$

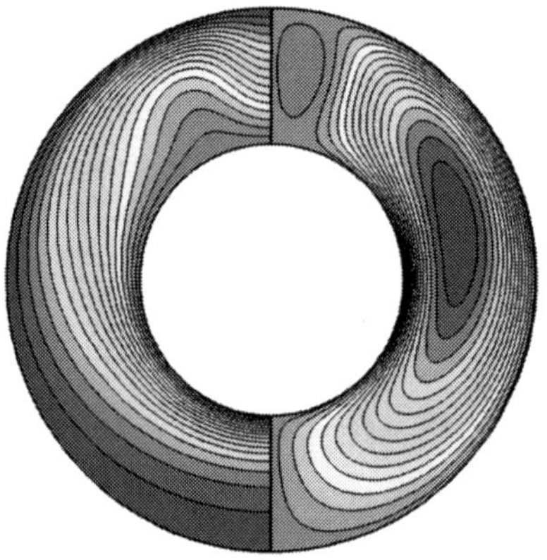

(b): $\eta = \sqrt{3}$

Fig. 5. Effect of η on the flow structure for $Ra = 63.77$

Numerical modelling of natural convection in a square cavity: effect of nanofluid volume fraction and inclination

M.M. Charafi[a], A. Bendaraa[b], A. Hasnaoui[c]

LS3M, Faculté Polydisciplinaire de Khouribga Univ. Sultan My Slimane, 25000 Khouribga, Morocco.

Abstract. In this paper, the phenomenon of natural convection in a square cavity filled with copper (Cu) - based nanofluid (Nf) has been studied. The influences of the volume fraction (ϕ) on the convective flow, as well as the effect of the inclination (γ) of the cavity on thermal performances were examined. The dimensionless governing equations formulated using stream function, vorticity, and temperatures were solved by finite difference methods, where the Successive Over-Relaxation (SOR) iteration and the upwind scheme were adopted. The developed code served as a verification of results where it showed a good agreement with previous works. The calculation code was then used to visualize the influence of the inclination on the thermal enhancement and showed that the transfer becomes more efficient with an inclination of $\pi/6$.

1 Introduction

The phenomenon of convection remains a current and much diversified research axis. This is justified by a set of recently published research works. The mastery and performance management of this type of transfer has become a major issue in the industry, its study has been the subject of several research projects crowned by publications in the field [1–5]. At the laboratory scale, several works have been focused on the study of convective flows in several geometries, namely square cavities.

Karatas et al. [6] carried out an experimental study of the phenomenon of convection in a rectangular cavity subjected to a radiation considered as a source of thermal excitation. In the same context, convection within a ventilated cavity has been the subject of a numerical work by Ismael el al. [7]. Thus, several numerical methods have been implemented to solve the problem of convection in cavities. Latice Boltzman Methode (LBM) has been used by Haouat et al. [8] to analyze the same problem. Wang et al. [9] have opted for Chebyshev spectral method to analyze convection in a cavity with a porous media.

Other works concentrated on the study of the nanofluids in the square cavities [10]. Boualit et al. [11] and Khanafer et al. [12] used finite element as well as finite volume methods to solve the same problems. Kogawa et al. [13] investigated the influence of the concentration of nanofluids on thermal convection.

The objective of this work is the development of a numerical code to study the phenomenon of convection in a square cavity filled with nanofluids. Thus, the simulation of the natural convection of the fluid under the effect of a temperature gradient, and the study of the effect of the nanoparticle volume fraction on the improvement of the heat transfer will allow the validation of the code, by comparing the present results with previous works. The code will then be used to analyze the effect of cavity inclination on thermal performance.

2 Mathematical formulation

The studied configuration is a square cavity with rigid walls, filled with a nanofluid. This configuration is subjected to a temperature gradient with the hot temperature (Th) imposed on the left wall while the cold temperature (Tc) is on the right wall (Dirichlet conditions). The other walls are adiabatic as it is shown in the schema presented in Fig.1.

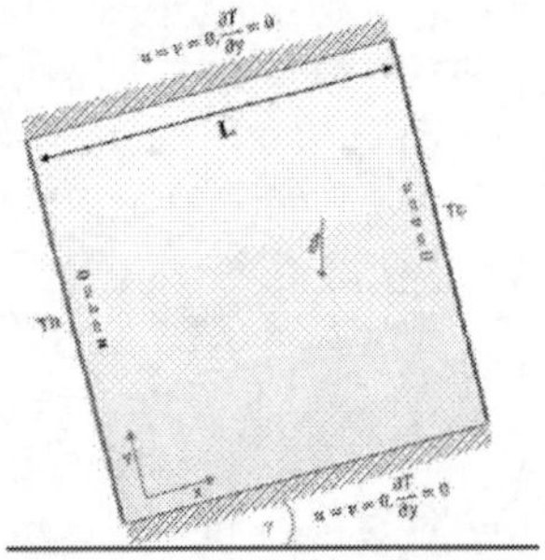

Fig. 1. Schematic representation for the physical model.

The used nanofluid is based on a suspension of copper (Cu) nanoparticles. The fluid flow in the cavity is assumed to be Newtonian and incompressible. The

[a] e-mail: mcharafi@hotmail.com
[b] e-mail: anass.bendaraa@gmail.com
[c] e-mail: hasnaoui59@hotmail.com

governing equations formulated using stream function, vorticity, and temperatures are expressed by the following system of equations:

$$\frac{\partial\omega}{\partial t}+u\frac{\partial\omega}{\partial x}+v\frac{\partial\omega}{\partial y}=\frac{\mu_{eff}}{\rho_{nf}}\left(\frac{\partial^2\omega}{\partial x^2}+\frac{\partial^2\omega}{\partial y^2}\right)+ \frac{g}{\rho_{nf}}\left[\phi\rho_s\beta_s+(1-\phi)\rho_f\beta_f\right]\left[\cos(\gamma)\frac{\partial T}{\partial x}-\sin(\gamma)\frac{\partial T}{\partial y}\right], \quad (1)$$

$$\frac{\partial T}{\partial t}+u\frac{\partial T}{\partial x}+v\frac{\partial T}{\partial y}=\frac{\partial}{\partial x}\left[(\alpha_{nf}+\frac{k_d}{(\rho c_p)_{nf}})\frac{\partial T}{\partial x}\right]+ \frac{\partial}{\partial y}\left[(\alpha_{nf}+\frac{k_d}{(\rho c_p)_{nf}})\frac{\partial T}{\partial y}\right], \quad (2)$$

$$\frac{\partial^2\psi}{\partial x^2}+\frac{\partial^2\psi}{\partial y^2}=-\omega \quad (3)$$

With the following initial and boundary conditions:

$$u=v=T=0 \quad at \quad t=0 \quad every\ where \quad (4)$$

$$\begin{cases} at\ t>0 \\ u=v=\frac{\partial T}{\partial y}=0 \quad for\ y=0,\mathrm{L}\ \text{and}\ 0\le x\le L \\ T=T_h,\ u=v=0 \quad for\ x=0 \quad \text{and}\ 0\le y\le L \\ T=T_c,\ u=v=0 \quad for\ y=L \quad \text{and}\ 0\le x\le L \end{cases} \quad (5)$$

Here, x and y are spatial coordinates, u and v are the velocities components, T is the temperature, α_{nf} the nanofluid thermal diffusivity, μ_{eff} is the effective dynamic viscosity, ρ_{nf}, ρ_s and ρ_f are respectively the densities of nanofluid, nanoparticles and the density of fluid, β_{nf}, β_f and β_s are respectively the thermal expansion of nanofluid fluid and nanoparticles, ϕ is the volume fraction, k_d is the thermal conductivity of the dispersion, and $(k_{eff})_{stagnant}$ is the effective stagnant conductivity [12], C_p the specific heat, the constant C is determined experimentally C=0.4 [11-12] for cooper-water nanofluid, ψ is the stream function and ω is the vorticity.

To obtain the dimensionless equations, we introduce the dimensionless spatial variables undermentioned:

$$X=\frac{x}{L},\ Y=\frac{y}{L},\ U=\frac{u}{\sqrt{g\beta_f\Delta T\,L}},\ V=\frac{v}{\sqrt{g\beta_f\Delta T\,L}}$$
$$\tau=\frac{t\sqrt{g\beta_f\Delta T\,L}}{L},\ \Omega=\frac{\omega L}{\sqrt{g\beta_f\Delta T\,L}},\ \Psi=\frac{\psi}{L\sqrt{g\beta_f\Delta T\,L}},\ \theta=\frac{T-T_c}{T_h-T_c}$$

The system of dimensionless equations will be written as follows:

$$\frac{\partial\Omega}{\partial\tau}+U\frac{\partial\Omega}{\partial X}+V\frac{\partial\Omega}{\partial Y}=\frac{\left(\frac{\partial^2\Omega}{\partial X^2}+\frac{\partial^2\Omega}{\partial Y^2}\right)}{(1-\phi)^{2.5}\left[\phi\frac{\rho_s}{\rho_f}+(1-\phi)\right]\sqrt{Gr}}+ \lambda\left[\cos(\gamma)\frac{\partial\theta}{\partial X}-\sin(\gamma)\frac{\partial\theta}{\partial Y}\right], \quad (6)$$

$$\frac{\partial\theta}{\partial\tau}+U\frac{\partial\theta}{\partial X}+V\frac{\partial\theta}{\partial Y}=\frac{1}{\Pr\sqrt{Gr}}\left[\frac{\partial}{\partial X}\left(\chi\frac{\partial\theta}{\partial X}\right)+\frac{\partial}{\partial Y}\left(\chi\frac{\partial\theta}{\partial Y}\right)\right], \quad (7)$$

$$\frac{\partial^2\Psi}{\partial X^2}+\frac{\partial^2\Psi}{\partial Y^2}=-\Omega \quad (8)$$

With:

$$\chi=\frac{\left[\frac{(k_{eff})_{stagnant}}{k_f}\right]}{(1-\phi)+\phi\frac{(\rho C_p)_s}{(\rho C_p)_f}}+C\phi\frac{d_p}{L}\Pr\sqrt{Gr}\sqrt{U^2+V^2} \quad (9)$$

And:

$$\frac{(k_{eff})_{stagnant}}{k_f}=\frac{k_s+2k_f-2\phi(k_f-k_s)}{k_s+2k_f+2\phi(k_f-k_s)} \quad (10)$$

In this work, the particle diameter used is d_p=10*nm*, and the characteristic dimension of the cavity "L" is taken equal to 1cm,

The dimensionless numbers are expressed using the thermophysical properties of the base fluid. The Prandtl number (Pr) and the Grashof number (Gr) as well as the coefficient λ which appears in the term of thermal expansion are given by:

$$\Pr={\nu_f}/{\alpha_f},\quad Gr={g\beta_f\Delta T\,L^3}/{\nu_f^2},\quad \lambda=\frac{\beta_{nf}}{\beta_f}$$

In order to quantify the thermal transfer within the cavity, and to characterize the effect of increasing nanofluid volume fraction, the value of the local Nusselt and the Nusselt average must be determined. These parameters are calculated as follows:

$$Nu=-\frac{k_{nf}}{k_f}\left.\frac{\partial\theta}{\partial X}\right|_{X=0},\quad Nu_{aver}=\int_0^1 Nu(Y)\,\mathrm{d}Y \quad (11)$$

3 Numerical method

The governing Eqs. (6)–(8) with corresponding initial and boundary conditions have been solved by finite difference method, where the Successive Over-Relaxation (SOR) iteration is used for the stream function equation (8) and the Upwind scheme is used for vorticity and temperature equation (6-7). The convergence criterion used is: $\frac{\sum\left|A_{i,j}^{n+1}-A_{i,j}^{n}\right|}{\sum\left|A_{i,j}^{n+1}\right|}<10^{-6}$ where the parameters A represents, the stream function Ψ, the temperature θ or the vorticity Ω.

Vorticities at the boundaries are found from Taylor's developments of the stream functions, and are be given as follows:

$$\begin{cases} \Omega_{i,1}=-2\frac{\Psi_{i,2}}{(\Delta X)^2},\ \Omega_{i,nx}=-2\frac{\Psi_{i,ny-1}}{(\Delta X)^2} \\ \Omega_{1,j}=-2\frac{\Psi_{2,j}}{(\Delta Y)^2},\ \Omega=-2\frac{\Psi_{nx-1,j}}{(\Delta Y)^2} \end{cases}$$

4 Volume fraction effects

To validate the developed code, a series of numerical experiments carried out to describe the flow profiles into the cavity with γ=0°. These experiments are done for several mesh spacing from 41×41, to 91×91. We choose the spacing 81×81 for his better ratio between simulation time and accuracy of the results. These results are compared with those of Khanafer et al. [12]. In addition to the good agreement with the results of Khanafer et al. [12], the influence of the volume fraction on the streamlines and isotherms is shown in Figs. 2-5. These figures illustrate the effects of the volume fraction on the isothermal lines and the stream lines for a number of Gr=10^4. The analysis of these figures shows that the increase of the nanoparticle volume fraction leads to an intensification of the central vortices, which tends to become elliptical. These results can be explained by the influence of the volume fraction of the nanoparticles on the viscous forces and the buoyant forces interacting during a natural convection flow.

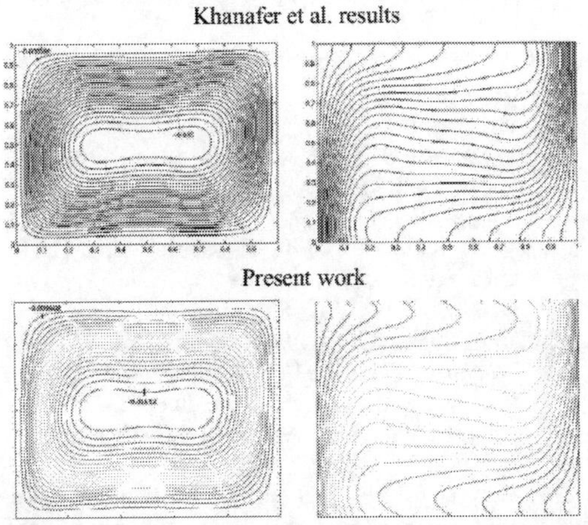

Fig. 2. Comparison of the streamlines and the isotherms between the present work and [12] (Pr=6.2, Gr=10^4, ϕ=0%, γ=0°).

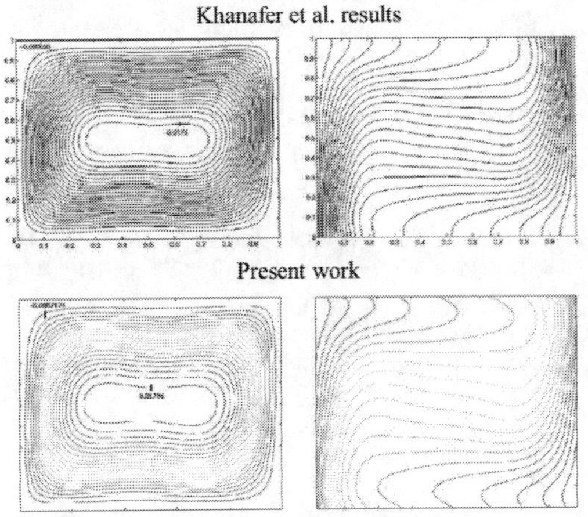

Fig. 3. Comparison of the streamlines and the isotherms between the present work and [12] (Pr=6.2, Gr=10^4, ϕ=8%, γ=0°).

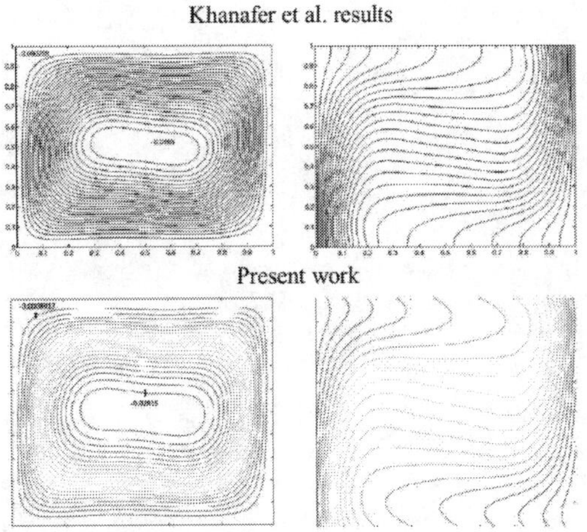

Fig. 4. Comparison of the streamlines and the isotherms between the present work and [12] (Pr=6.2, Gr=10^4, ϕ=16%, γ=0°).

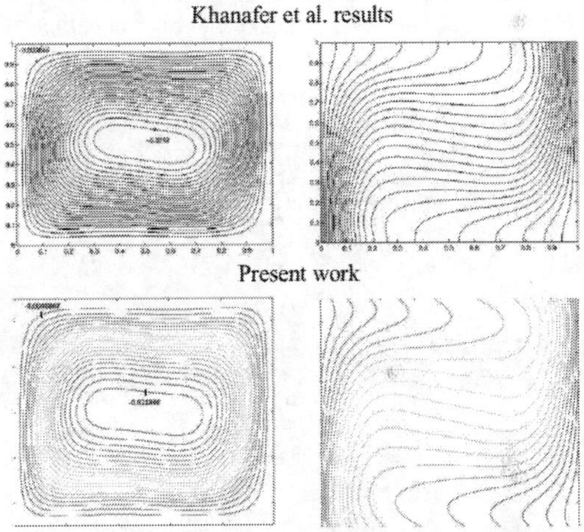

Fig. 5. Comparison of the streamlines and the isotherms between the present work and [12] (Pr=6.2, Gr=10^4, ϕ=20%, γ=0°).

A comparison of the Nusselt values, between the present work and Khanafer et al. [12] is presented in Fig. 6-a. This figure shows a good agreement between the results. It can be also seen that the transfer intensifies by increasing the volume fraction that is explained by the **Nu** increasing.

5 Inclination effects

In this section, after testing the calculation code, which gave a good agreement with result of Khanafer et al. [12], the influence of the inclination of the cavity on the nanofluid flow, as well as, on the thermal performances are investigated, where a series of simulations are performed. A cooper (Cu) nanoparticle volume fraction of 10% is used with different inclinations ranging from 0° to $\pi/2$ and for values of Grashof number ranging from 10^3 to 10^5.

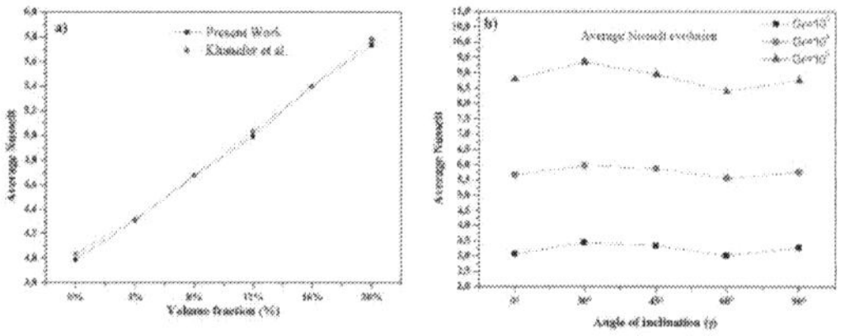

Fig. 6. a) The comparison of Nusselt between the present work & Khanafer (Pr=6.2, γ=0°). **b)** Evolution of the Nusselt number as a function of the inclination angle for different **Gr** numbers (Pr=6.2, ϕ=0%).

Figure 6-b illustrates the evolution of the average number of Nusselt in the vicinity of the hot wall, thus allowing quantifying the rate of improvement of the heat transfer by convection in the cavity. As shown in Fig. 6, the heat transfer becomes more intense by increasing the number of Grashof (Gr). On the other hand, the inclination of the cavity has an influence on the number of Nusselt and marks a maximum for an inclination of $\pi/6$ and a minimum of performance at an inclination of $\pi/3$.

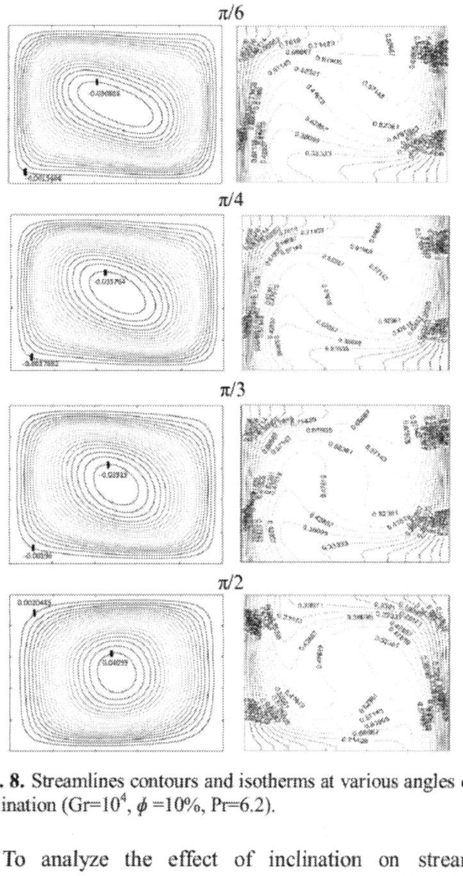

Fig. 8. Streamlines contours and isotherms at various angles of inclination (Gr=10^4, ϕ =10%, Pr=6.2).

To analyze the effect of inclination on stream function and temperatures, the case of a nanofluid with a volume fraction of 10% was adopted. Fig.7 shows that increasing the angle of inclination produces a change in the stream function and temperature profiles. Indeed, we can notice that the increase of the inclination has a direct effect on the intensification of the contours and on the profiles of temperature, where we can notice the migration of the temperature isolines from the hot source towards the cold source

6 Conclusion

In this paper, a code based on a finite difference method using the S.O.R iteration and the upwind system was developed. Several numerical simulations have been performed to validate the code, comparing the results with previous works. The good agreement of the results of the code was thus confirmed, which allowed using this code as a prediction tool in the study of the influence of the nanofluid concentration as well as the inclination of the cavity on the thermal performances. An analysis of the results concluded that the improvement in heat transfer is proportional to the increase in the volume fraction of the nanofluid, and that the variation of the inclination affects the thermal enhancement, where the transfer becomes more efficient with tilt of $\pi/6$.

References

1. K. Fouladi, A. P. Wemhoff, L. Silva -Llanca, K. Abbasi, and A. Ortega, *Appl. Therm. Eng.*, vol. **124**, pp. 929–939, (2017).
2. A. R. Rahmati, A. Rayat Roknabadi, and M. Abbaszadeh, *Alexandria Eng. J.*, vol. **55**, no. 4, pp. 3101–3114, (2016).
3. A. Y. Snegirev, *Combust. Flame*, vol. **136**, no. 1–2, pp. 51–71, (2004).
4. A. S. Yang, C. Y. Wen, Y. H. Juan, Y. M. Su, and J. H. Wu, *Appl. Therm. Eng.*, vol. **70**, no. 1, pp. 219–230, (2014).
5. A. E. Conradie, J. D. Buys, and D. G. Kröger, *Appl. Therm. Eng.*, vol. **18**, no. 97, pp. 25–45, (1998).
6. H. Karatas and T. Derbentli, *Int. J. Therm. Sci.*, vol. **123**, pp. 129–139, (2018).
7. M. A. Ismael and H. F. Jasim, *Int. J. Mech. Sci.*, vol. **135**, pp. 190–202, (2018).
8. S. Houat and Z. E. Bouayed, *Energy Procedia*, vol. **139**, pp. 186–191, (2017).
9. Y. Wang, G. Qin, W. He, and Z. Bao , *Int. J. Heat Mass Transf.*, vol. **121**, pp. 1055–1072, (2018).
10. T. Armaghani, A. Kasaeipoor, N. Alavi, and M. M. Rashidi, *J. Mol. Liq.*, vol. **223**, pp. 243–251, (2016).
11. A. Boualit, N. Zeraibi, T. Chergui, M. Lebbi, L. Boutina, and S. Laouar, *Int. J. Hydrogen Energy*, vol. **42**, no. 13, pp. 8611–8623, (2017).
12. K. Khanafer, K. Vafai, and M. Lightstone, *Int. J. Heat Mass Transf.*, vol. **46**, no. 19, pp. 3639–3653, (2003).
13. T. Kogawa, L. Chen, J. Okajima, A. Sakurai, A. Komiya, and S. Maruyama, *Appl. Therm. Eng.*, vol. **131**, pp. 141–149, (2018).

Global stability analysis of a delayed HIV model with saturated infection rate

jaouad Danane [a] and Karam Allali [b]

Laboratory of Mathematics and Applications,
Faculty of Sciences and Technologies, University Hassan II
of Casablanca, PO Box 146, Mohammedia,Morocco

Abstract. In this paper, the global stability of a delayed HIV model with saturated infection rate infection is investigated. We incorporate two discrete delays into the model; the first describes the intracellular delay in the production of the infected cells, while the second describes the needed time for virions production. We also derive the global properties of this two-delay model as function of the basic reproduction number R_0. By using some suitable Lyapunov functions, it is proved that the free-equilibrium point is globally asymptotically stable when $R_0 \leq 1$, and the endemic equilibrium point is globally asymptotically stable when $R_0 \geq 1$. Finally, in order to support our theoretical findings we have illustrate some numerical simulations.

Keywords: delay model, HIV infection, viral dynamics, global stability, numerical simulation.

1 Introduction

The human immunodeficiency virus (HIV) is a virus that gradually weakens the immune system. It is considered as the main cause for several deadly diseases after the resulting acquired immunodeficiency syndrome (AIDS) is reached. With 36.7 million people living with HIV, 2.1 million people becoming newly infected by HIV and more than 1.1 million deaths annually, HIV becomes a major global public health issue [1]. For this reason, several mathematical models describing the dynamics of HIV infection have been developed and studied [2–6]. The basic of them was suggested in [2]:

$$\begin{cases} \dot{x} = \lambda - d_1 x(t) - k_1 x(t) v(t), \\ \dot{y} = k_1 x(t) v(t) - d_2 y(t), \\ \dot{v} = a y(t) - d_3 v(t), \end{cases} \tag{1}$$

where x, y and v denote the concentration of uninfected CD4$^+$ T cells, infected CD4$^+$ T cells and free virus, respectively. Susceptible host cells CD4$^+$ T cells are produced at a rate λ, die at a rate d_1 and become infected by virus at a rate $k_1 xv$. Infected cells die at a rate d_2. Finally, the free virus is produced by infected cells at a rate a and decays at a rate d_3. In our model we propose to change the mass action rate in (1) by a saturated infection rate $\dfrac{k_1 x(t) v(t)}{x(t)+v(t)}$, called the saturated mass action [7] which describes better the rate of viral infection, then the viral infection model becomes

$$\begin{cases} \dot{x}(t) = \lambda - d_1 x(t) - \dfrac{k_1 x(t) v(t)}{x(t)+v(t)}, \\ \dot{y}(t) = \dfrac{k_1 x(t) v(t)}{x(t)+v(t)} - d_2 y(t) \\ \dot{v}(t) = a y(t) - d_3 v(t), \end{cases} \tag{2}$$

with the initial conditions $x(0) = x_0$, $y(0) = y_0$ and $v(0) = v_0$. The theoretical analysis of the epidemic models [9–12] show that the time delay plays a crucial role in the dynamical properties of the HIV infection models, in order to study the influence of time delay on the spread of HIV, we have illustrated the following model:

$$\begin{cases} \dot{x}(t) = \lambda - d_1 x(t) - \dfrac{k_1 x(t) v(t)}{x(t)+v(t)}, \\ \dot{y}(t) = \dfrac{k_1 x(t-\tau_1) v(t-\tau_1)}{x(t-\tau_1)+v(t-\tau_1)} - d_2 y(t) \\ \dot{v}(t) = a y(t-\tau_2) - d_3 v(t). \end{cases} \tag{3}$$

In this model, x, y and v denote the concentration of uninfected cells, infected cells and free virus respectively. Where τ_1 represents the delay in the production of productively infected cells and τ_2 represents the delay in the production of virus. With the initial conditions of model (3) are given by $x(\theta) > 0$, $y(\theta) > 0$ and $v(\theta) > 0$ for $\theta \in [-\rho, 0]$, where $\rho = max(\tau_1, \tau_2)$.

[a] e-mail: `jaouaddanane@gmail.com`
[b] e-mail: `allali@hotmail.com`

2 Global analysis of the delay model

2.1 Steady states

For our model (2), the basic reproduction number of the virus is given by

$$R_0 = \frac{ak_1}{d_2 d_3}. \quad (4)$$

The model (2) has two steady states:

- The disease-free steady state $E_f = (\frac{\lambda}{d_1}, 0, 0)$.
- The endemic steady state

$$E^* = (x_1, y_1, v_1),$$

where

$$x_1 = \frac{\lambda}{d_1 + k_1(1 - \frac{1}{R_0})}, y_1 = \frac{\lambda k_1(1 - \frac{1}{R_0})}{d_2(d_1 + k_1(1 - \frac{1}{R_0}))}$$

and $v_1 = \dfrac{\lambda(R_0 - 1)}{d_1 + k_1(1 - \frac{1}{R_0})}$.

2.2 Global analysis

In this subsection, we study the global asymptotic stability of the steady states using a suitable Lyapunov functional. For this reason, we consider the function

$$F(x) = x - 1 - ln(x) \forall x > 0.$$

Note that $F(x) > 0, \forall x > 0$ and that $F(x) = 0$ if and only if $x = 1$. We consider the positive real numbers $x_1, x_2, ..., x_n$, then

$$1 - x_i + \ln x_i = -F(x_i) \leq 0, \quad \forall i = 1, ..., n \quad (5)$$

By summing for $i = 1, ..., n$, we get:

$$n - \sum_{i=1}^{n} x_i + \ln \prod_{i=1}^{n} x_i \leq 0, \quad (6)$$

if we take $x_i = \frac{p_i}{q_i}$, we will have

$$n - \sum_{i=1}^{n} \frac{p_i}{q_i} + \ln \prod_{i=1}^{n} \frac{p_i}{q_i} \leq 0, \quad (7)$$

Theorem 1 *If $R_0 \leq 1$, then the disease-free steady state E_f is globally asymptotically stable for any delays $\tau_1 > 0$ and $\tau_2 > 0$.*

Proof. Let the following Lyapunov functional:

$$\mathcal{L}_1(x, y, v) = y + \frac{d_2}{a} v + k_1 \int_{t-\tau_1}^{t} \frac{x(\xi)v(\xi)}{x(\xi) + v(\xi)} d\xi + d_2 \int_{t-\tau_2}^{t} y(\xi) d\xi.$$

The time derivative is given by:

$$\begin{aligned}\dot{\mathcal{L}}_1(x, y, v) &= \dot{y} + \frac{d_2}{a}\dot{v} \\ &+ k_1 \left(\frac{x(t)v(t)}{x(t) + v(t)} - \frac{x(t-\tau_1)v(t-\tau_1)}{x(t-\tau_1) + v(t-\tau_1)} \right) \\ &+ d_2 (y(t) - y(t-\tau_2)) \\ &= \frac{k_1 x(t)v(t)}{x(t) + v(t)} - \frac{d_2 d_3}{a} v(t) \\ &\leq k_1 v(t) - \frac{d_2 d_3}{a} v(t) \\ &\leq \frac{d_2 d_3}{a} (R_0 - 1) v(t).\end{aligned}$$

If $R_0 < 1$; then $\dot{\mathcal{L}} \leq 0$. Moreover, $\dot{\mathcal{L}} \leq 0$ when $v = 0$. The largest compact invariant is

$$E = \{(x, y, v) \in R_*^+ | v = 0\}.$$

According to LaSalle's invariance principle, $\lim_{+\infty} v(t) = 0$, the limit system of equations is:

$$\begin{cases} \dot{x} = \lambda - d_1 x \\ \dot{y} = -d_2 y \end{cases}$$

We define:

$$\mathcal{L}_1(x, y) = \frac{1}{x_0}\left(x - x_0 - x_0 \ln \frac{x}{x_0}\right) + y.$$

Since $x_0 = \frac{\lambda}{d_1}$, then

$$\dot{\mathcal{L}}(x, y) = d_1 \left(2 - \frac{x}{x_0} - \frac{x_0}{x}\right) - d_2 y.$$

Since the arithmetic mean is greater than or equal to the geometric mean, it follows

$$2 - \frac{x}{x_0} - \frac{x_0}{x} \leq 0$$

Therefore $\dot{\mathcal{L}} \leq 0$ and the equality holds if $x = x_0$ and $y = 0$, which complete the proof.

Theorem 2 *The endemic steady state E^* is globally asymptotically stable when $R_0 > 1$, for any delays $\tau_1 > 0$ and $\tau_2 > 0$.*

Proof.
Let the following Lyapunov functional:

$$\begin{aligned}\mathcal{L}_2(x, y, v) &= x - x_1 - \int_{x_1}^{x} \frac{d_2 y_1}{\frac{k_1 u v_1}{u + v_1}} du + y - y_1 \\ &- y_1 \ln \frac{y}{y_1} + \frac{d_2}{a}(v - v_1 - v_1 \ln \frac{v}{v_1}) \\ &+ k_1 \frac{x_1 v_1}{x_1 + v_1} \int_{t-\tau_1}^{t} F(\frac{x(\xi)v(\xi)}{x_1 v_1} \frac{x_1 + v_1}{x(\xi) + v(\xi)}) d\xi \\ &+ d_2 y_1 \int_{t-\tau_2}^{t} F(\frac{y(\xi)}{y_1}) d\xi.\end{aligned}$$

then, we have

$$\begin{aligned}\dot{\mathcal{L}}_2(x,y,v) &= -\frac{d_1 v_1}{x(x_1+v_1)}(x-x_1)^2 \\ &- (d_2+k_2)y_1\left(\frac{x(v-v_1)^2}{v_1(x+v_1)(x+v)}\right) \\ &+ d_2 y_1\Bigg(4 - \frac{x_1}{x}\frac{x+v_1}{x_1+v_1} - \frac{y_1}{y}\frac{x(t-\tau_1)v(t-\tau_1)}{x_1 v_1} \\ &\times \frac{x_1+v_1}{x(t-\tau_1)+v(t-\tau_1)} - \frac{y(t-\tau_2)v_1}{y_1 v} - \frac{x+v}{x+v_1} \\ &+ \ln\bigg(\frac{x(t-\tau_1)+v(t-\tau_1)}{xv}\frac{x+v}{x(t-\tau_1)+v(t-\tau_1)} \\ &\frac{y(t-\tau_2)}{y}\bigg)\Bigg),\end{aligned}$$

Taking $p_1 = x_1(x+v_1)$, $p_2 = y_1(x(t-\tau_1)v(t-\tau_1))$, $p_3 = y(t-\tau_2)v_1$, $p_4 = x+v$, $q_1 = x(x_1+v_1)$, $q_2 = y(x_1 v_1)$, $q_3 = y_1 v$, $q_4 = x+v_1$ and using the inequality (7) for $n = 4$, we obtain

$$\begin{aligned}&4 - \frac{x_1}{x}\frac{x+v_1}{x_1+v_1} - \frac{y_1}{y}\frac{x(t-\tau_1)v(t-\tau_1)}{x_1 v_1} \\ &\times \frac{x_1+v_1}{x(t-\tau_1)+v(t-\tau_1)} - \frac{y(t-\tau_2)v_1}{y_1 v} - \frac{x+v}{x+v_1} \\ &+ \ln\left(\frac{x(t-\tau_1)+v(t-\tau_1)}{xv}\frac{x+v}{x(t-\tau_1)+v(t-\tau_1)}\right. \\ &\left.\frac{y(t-\tau_2)}{y}\right) \le 0\end{aligned}$$

Therefore, $\dot{\mathcal{L}}_2 \le 0$. Assume that M is the largest invariant set in $(x,y,v)|\dot{\mathcal{L}}_2 = 0$. Note that $\dot{\mathcal{L}}_2 = 0$ if and only if $x = x_1$, $y = y_1$ and $v = v_1$. Hence, $M = E^*$. Since E^* exists whenever $R_0 > 1$, then by the Lyapunov-LaSalle invariance theorem E^* is globally asymptotically stable if $R_0 > 1$.

3 Numerical results and simulations

In order to perform the numerical simulations, the system (1.1) will be solved numerically using the fourth order Runge-Kutta iterative scheme. The used parameter values are taken from Table 1.

Figure 1 represents the uninfected cells, the infected cells and the virus as function of time. The solid curves represent the case of $(\tau_1,\tau_2) = (5,5), (10,10)$ and $(30,30)$. For the parameters used in this figure, the basic reproduction number is $R_0 = 0.39 < 1$; we clearly see that the delays do not affect the global asymptotic stability behavior of the disease-free equilibrium point $E_f = (719.42, 0, 0)$. This numerical result is consistent with the theoretical result concerning the stability of E_f. Finally, Fig. 2 represents the uninfected cells, the infected cells and the virus as function of time. The solid curves represent the case of $(\tau_1,\tau_2) = (5,5), (10,10)$ and $(30,30)$). For the parameters used in this figure, the basic reproduction number is $R_0 = 19.98 > 1$; we clearly see that the magnitude of the two delays does not affect the global asymptotic stability of the endemic equilibrium point $E^* = (214.49, 172.09, 2.97\times 10^4, 0, 0)$. This numerical result is consistent with the theoretical result concerning the stability of E^*.

Table 1. Parameters, their symbols and default values used in the suggested HIV model

Parameters	Meaning	Value	References
λ	Source rate of CD4+ T cells	$[0, 10]$	[5]
k_1	Average of infection	$[2.5\times 10^{-4}, 0.5]$	[3]
d_1	Decay rate of healthy cells	0.0139	[3]
d_2	Death rate of infected CD4+ T cells	0.5776	[3]
a	The rate of production the virus by infected CD4+ T cells	$[2, 1250]$	[3]
d_3	Clearance rate of virus	$[0.3466, 2.4]$	[3]

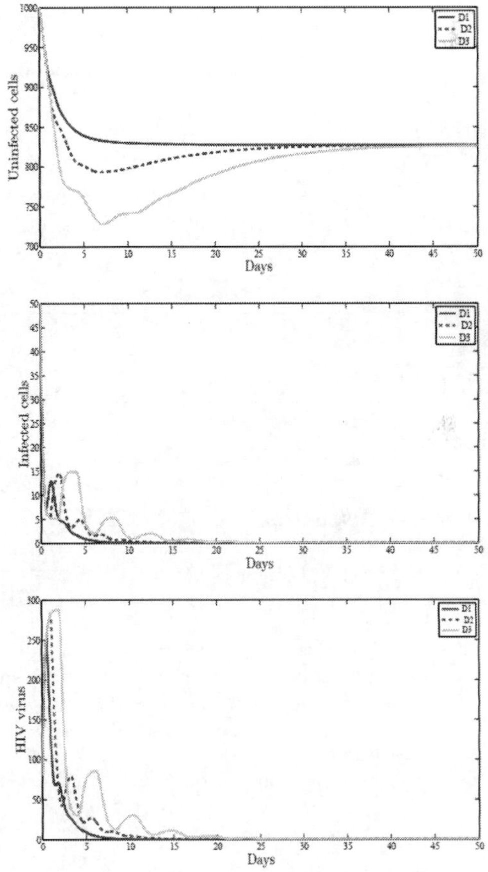

Fig. 1. The behavior of the infection dynamics for $\lambda = 10$, $d_1 = 0.0139$, $k_1 = 0.04$, $d_2 = 0.5776$, $a = 2$, $d_3 = 0.3466$ and $D_1 : (\tau_1,\tau_2) = (5,5); D_2 : (\tau_1,\tau_2) = (10,10); D_3 : (\tau_1,\tau_2) = (30,30)$.

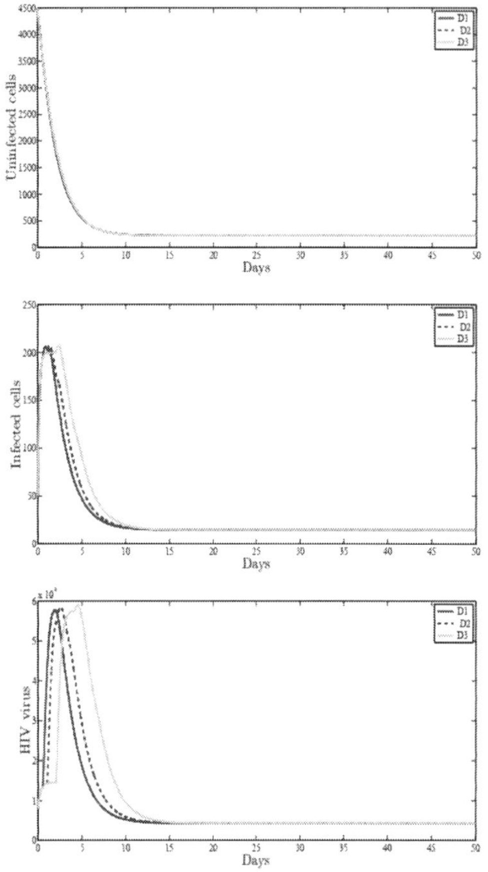

Fig. 2. The behavior of the infection dynamics for $\lambda = 10$, $d_1 = 0.0139$, $k_1 = 0.04$, $d_2 = 0.5776$, $a = 100$, $d_3 = 0.3466$ and $D_1 : (\tau_1, \tau_2) = (5, 5); D_2 : (\tau_1, \tau_2) = (10, 10); D_3 : (\tau_1, \tau_2) = (30, 30)$.

4 Discussion

The HIV dynamics involving three densities of the uninfected cells, the infected cells and the HIV virus have been widely studied in different mathematical models [2–6]. In this work, we propose a modified viral dynamics model, using a saturated incidence function with two delays in the generation of infected cells and in the production of viruses. It was observed that the delays have a clear effect on the dynamics of the HIV infection by slowing down viral replication and reducing the amount of the infected cells.

5 Conclusion

In this work, we have studied the model of the HIV dynamics with time delay. The model includes a saturated rates in order to better describe the viral infection. Further, we have incorporated two delays in the model where the first delay stands for the needed time of the production of infected cells from the uninfected ones and the second delay describes the needed time for the virus production. These two delays have been included in the model to check if these delays can affect the global stability. The results show that inclusion of these delays does not influence the global dynamics of the delay model. Finally, the numerical simulations are performed in order to show the behavior of infection during the days of observation and to support the theoretical findings.

References

1. World Health Organization HIV/AIDS Key facts, (July 2014), http://www.who.int.
2. Nowak, M.A., Bonhoeffer, S., Shaw, G.M., May, R.M.(1997) Anti-viral drug treatment: Dynamics of resistance in free virus and infected cell populations. J. Theor. Biol. 184(2):203-217.
3. Sun, Q., Min, L., & Kuang, Y. (2015) Global stability of infection-free state and endemic infection state of a modified human immunodeficiency virus infection model. IET systems biology, 9(3), 95-103.
4. Allali, K., Danane, J. and Kuang, Y., 2017. Global Analysis for an HIV Infection Model with CTL Immune Response and Infected Cells in Eclipse Phase. Applied Sciences (2076-3417), 7(8).
5. Wang, Y., Zhou, Y., Wu, J., Heffernan, J.(2009). Oscillatory viral dynamics in a delayed HIV pathogenesis model. Math. Biosci. 219, 104-112.
6. Wang, X., Wang, W. (2012) An HIV infection model based on a vectored immunoprophylaxis experiment. Journal of theoretical biology 313:127-135.
7. Q. Sun, L. Min. Dynamics Analysis and Simulation of a Modified HIV Infection Model with a Saturated Infection Rate. *Computational and mathematical methods in medicine*, 2014; **2014**, Article ID 145162, 14 pages.
8. Hale J, Verduyn Lunel SM(1993) Introduction to functional differential equations, applied mathematical science, vol 99. Springer, New York.
9. Y. Kuang, Delay Differential Equations with Applications in Population Dynamics, Academic Press, San Diego, 1993.
10. R.V. Culshaw, S. Ruan, A delay-differential equation model of HIV infection of D4+ T-cells, Math. Biosci. 165(2000) 27-39.
11. R.V. Culshaw, S. Ruan, G. Webb, A mathematical model of cell-to-cell HIV-1 that include a time delay, J. Math.Biol. 46 (2003) 425-444.
12. J.E. Mittler, B. Markowitz, D.D. Ho, A.S. Perelson, Improved estimates for HIV-1 clearance rate and intracellular delay, AIDS 13 (1999) 1415-1417.

Programmed task based motion analysis of robotic systems equipped with flexible links and supports

Elżbieta Jarzębowska[1*], Krzysztof Augustynek[2], and Andrzej Urbaś[2]

[1]*Warsaw University of Technology, 00-665 Warsaw, Nowowiejska 24, Poland*
[2]*University of Bielsko-Biala, 43-309 Bielsko-Biala, Willowa 2, Poland*

Abstract. The automatic computational procedure to derivate dynamics equations of systems with programmed constraints modified to encompass compliant mechanical components in their structures is discussed in the paper. The dynamics analysis of the compliant manipulator model with a flexible link is presented as an example. The rigid finite element method is used in order to take into account the flexibility of the link. The formalism of joint coordinates and homogeneous transformations are used to describe the manipulator motion.

1 Introduction

Analysis of robot task based motions dedicated to work and services delivery is presented in the paper. The key contribution of the paper is in three aspects of the analysis. The first one is the possibility of constraints incorporation into the system dynamics, the second one is the special computational procedure for the constrained dynamics generation, which provides dynamic models, referred to as reference dynamics, satisfying all constraints upon them. The third contribution comparing to the results reported in the literature, see e.g. [1, 2] is in modeling flexibility of the system parts and supports. The constraints put on a system are referred to as programmed and they are imposed as control goals on system performance or service task requirements. The procedure of generating reference dynamics offers automated generation of equations and it was successfully developed and implemented to rigid system models [3, 4]. The advantage of this procedure is that it serves both reference and control oriented dynamics derivation and the final dynamics models are obtained in the reduced state form, i.e. constraint reaction forces are eliminated. The procedure is extended on flexible subsystems of lightweight and fast machine parts and servicing equipment, which are prone to vibration in some work regimes. Vibrations may significantly affect system performance and disable effective controller designs. Flexibility of links is modeled using the rigid finite element method, whose advantage relies in its ability of application of the rigid body approach to modeling flexible link elements. The novelty of the presented method is its ability to analyze any system reference motion, including flexible link vibration, in the presence of kinematic task-based constraints. The results of this analysis can contribute to verification of a system behavior when it is subjected to given kinematic constraints, help to specify desired task-based constraints properly, e.g. put tighter velocity limits, to exclude some work regimes or to design controllers correctly. The possibility of reference motion analysis in these aspects enables simulation of various work regimes for system models and selection of required and safe task based motion parameters, accordingly. The theoretical development is illustrated by simulation studies of an example of a flexible link and support manipulator model, whose service tasks are predefined by the programmed constraints. Special interest of this study is to analyze flexible link vibrations when the manipulator performs the tasks.

2 Mathematical model of a manipulator

Figure 1 presents a compliant supported manipulator model. It is assumed, that the manipulator is driven by means of a driving torque $\mathbf{t}_{dr}^{(2)}$. The manipulator third link is flexible. The rigid finite element method is used to discretize this link.

The following joint coordinates are used to describe motion of the manipulator:

$$\mathbf{q} = (q_i)_{i=1,\dots,n_{dof}} = \left[\tilde{\mathbf{q}}^{(1)^T} \mid \tilde{\mathbf{q}}^{(2)^T} \mid \tilde{\mathbf{q}}^{(3)^T} \mid \tilde{\mathbf{q}}^{(4)^T} \right]^T, \tag{1}$$

where: $\tilde{\mathbf{q}}^{(1)} = \left[z^{(1)} \right]$, $\tilde{\mathbf{q}}^{(2)} = \left[\psi^{(2)} \right]$,

* Corresponding author: elajarz@meil.pw.edu.pl

https://doi.org/10.1051/matecconf/201824101008

$$\tilde{\mathbf{q}}^{(3)} = \left[\tilde{\mathbf{q}}^{(3,0)^T} \ \cdots \ \tilde{\mathbf{q}}^{(3,r)^T} \ \cdots \ \tilde{\mathbf{q}}^{(3,n_{rfe}^{(3)})^T}\right]^T,$$

$$\tilde{\mathbf{q}}^{(3,0)} = \left[\psi^{(3,0)}\right],\ \tilde{\mathbf{q}}^{(3,r)} = \left[\psi^{(3,r)} \ \ \theta^{(3,r)} \ \ \varphi^{(3,r)}\right]^T,$$

$$\tilde{\mathbf{q}}^{(4)} = \left[z^{(4)}\right].$$

Vector $\mathbf{q}$ contains independent and dependent coordinates:

$$\mathbf{q} = \left(q_i\right)_{i\in i_{i_c}\cup i_{d_c}} \in \left\{\mathbf{q}_{i_c}, \mathbf{q}_{d_c}\right\}, \tag{2}$$

where: $i_{i_c} \in \left\{1,2,4,\ldots,n_{dof-1}\right\}$ and $i_{d_c} \in \left\{3,\ldots,n_{dof}\right\}$ are independent and dependent coordinates indexes.

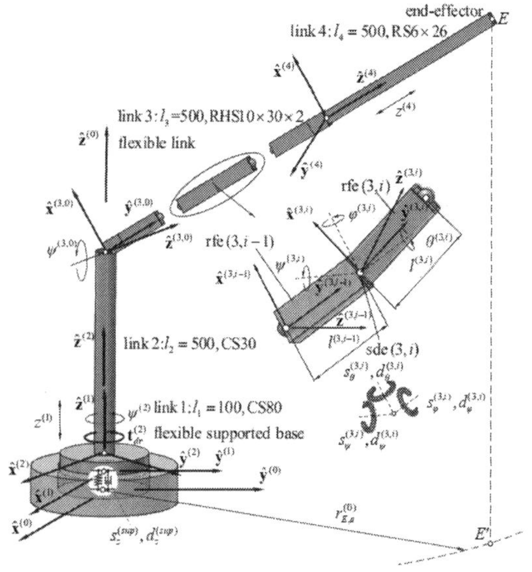

Figure 1. Model of a manipulator.

The generalized programmed motions equations (GPME) with first order programmed constraints underlie the derivation procedure and they have the form [3, 4]:

$$\left.\frac{\partial R_1}{\partial \dot{q}_i}\right|_{i\in i_{i_c}} + \sum_{j\in i_{d_c}} \frac{\partial R_1}{\partial \dot{q}_j}\frac{\partial \dot{q}_j}{\partial \dot{q}_i} = 0, \tag{3}$$

where:

$$R_1 = \dot{E}_k - 2\sum_{i=1}^{n_{dof}} \frac{\partial E_k}{\partial q_i}\dot{q}_i + \sum_{i=1}^{n_{dof}} \frac{\partial E_p}{\partial q_i}\dot{q}_i + \sum_{i=1}^{n_{dof}} \frac{\partial R_s}{\partial \dot{q}_i}\dot{q}_i - \sum_{i=1}^{n_{dof}} t_i\,\dot{q}_i,$$

$$E_k = \sum_{l=1}^{n_l} E_k^{(l)},\ E_k^{(l)} = \frac{1}{2}\mathrm{tr}\left\{\dot{\mathbf{T}}^{(l)}\mathbf{H}^{(l)}\left(\dot{\mathbf{T}}^{(l)}\right)^T\right\},$$

$$E_p = E_{p,g} + E_{p,sup} + E_{p,f_l},\ E_{p,g} = \sum_{l=1}^{n_l} E_{p,g}^{(l)},$$

$$E_{p,g}^{(l)} = m^{(l)} g \mathbf{J}_3 \mathbf{T}^{(l)} \mathbf{r}_{C^{(l)}}^{(l)},\ \mathbf{J} = \begin{bmatrix}\mathbf{J}_1\\ \mathbf{J}_2\\ \mathbf{J}_3\end{bmatrix} = \begin{bmatrix}1 & 0 & 0 & 0\\ 0 & 1 & 0 & 0\\ 0 & 0 & 1 & 0\end{bmatrix},$$

$$E_{p,sup} = \frac{1}{2} s_z^{(sup)}\left(z^{(1)}\right)^2,$$

$$E_{p,f_l} = \frac{1}{2}\sum_{s=1}^{n_{se}^{(l)}}\left(\tilde{\mathbf{q}}^{(l,s)}\right)^T \mathbf{S}^{(l,s)}\tilde{\mathbf{q}}^{(l,s)},\ \mathbf{S}^{(l,s)} = \mathrm{diag}\left\{s_\psi^{(l,s)}, s_\theta^{(l,s)}, s_\varphi^{(l,s)}\right\},$$

$$R_s = R_{sup} + R_{f_l}, R_{sup} = \frac{1}{2} d_z^{(sup)}\left(\dot{z}^{(1)}\right)^2,$$

$$R_{f_l} = \frac{1}{2}\sum_{s=1}^{n_{se}^{(l)}}\left(\dot{\tilde{\mathbf{q}}}^{(l,s)}\right)^T \mathbf{D}^{(l,s)}\dot{\tilde{\mathbf{q}}}^{(l,s)},\ \mathbf{D}^{(l,s)} = \mathrm{diag}\left\{d_\psi^{(l,s)}, d_\theta^{(l,s)}, d_\varphi^{(l,s)}\right\},$$

$$\mathbf{t} = \left[0 \ \ t_{dr}^{(2)} \ \ \mathbf{0}\right]^T.$$

It is assumed that the end-effector moves along an elliptical trajectory designed in a plane parallel to the $x^{(0)}y^{(0)}$ plane and its $z_E^{(0)}$ coordinate has to change according to assumed time-dependent function $z_{E,a}^{(0)}(t)$. Thus, these programmed constraints can be written in the form:

$$\Phi_1 \equiv 0 \Rightarrow \left(\frac{x_E^{(0)}}{a_{E,a}^{(0)}}\right)^2 + \left(\frac{y_E^{(0)}}{b_{E,a}^{(0)}}\right)^2 - 1 = 0 \tag{4.1}$$

$$\Phi_2 \equiv 0 \Rightarrow z_E^{(0)} - z_{E,a}^{(0)}(t) = 0 \tag{4.2}$$

where: $x_E^{(0)} = \mathbf{J}_1\mathbf{T}^{(4)}\mathbf{r}_E^{(4)}$ $y_E^{(0)} = \mathbf{J}_2\mathbf{T}^{(4)}\mathbf{r}_E^{(4)}$ $z_E^{(0)} = \mathbf{J}_3\mathbf{T}^{(4)}\mathbf{r}_E^{(4)}$,

According to the GPME algorithm [3], the constraint equations are differentiated to obtain

$$\dot{\Phi}_1 \equiv 0 \Rightarrow \mathbf{u}\dot{\mathbf{q}} = \mathbf{0}, \tag{5.1}$$

$$\dot{\Phi}_2 \equiv 0 \Rightarrow \mathbf{C}_3\dot{\mathbf{q}} - \dot{z}_{E,a}^{(0)}(t) = 0, \tag{5.2}$$

$$\ddot{\Phi}_1 \equiv 0 \Rightarrow \mathbf{u}\ddot{\mathbf{q}} + v = 0, \tag{5.3}$$

$$\ddot{\Phi}_2 \equiv 0 \Rightarrow \mathbf{C}_3\ddot{\mathbf{q}} + d_3 - \ddot{z}_{E,a}^{(0)}(t) = 0, \tag{5.4}$$

with:

$$\mathbf{C} = \begin{bmatrix}\mathbf{C}_1\\ \mathbf{C}_2\\ \mathbf{C}_3\end{bmatrix} = \left(c_{ij}\right)_{\substack{i=1,2,3\\ j=1,\ldots,4}} = \mathbf{J}\left[\mathbf{T}_1^{(4)}\mathbf{r}_E^{(4)} \ \cdots \ \mathbf{T}_4^{(4)}\mathbf{r}_E^{(4)}\right],$$

$$\mathbf{d} = \left(d_i\right)_{i=1,\ldots,3} = \mathbf{J}\left(\left(\sum_{i=1}^{n_{dof}}\sum_{j=1}^{n_{dof}} \mathbf{T}_{ij}^{(4)}\dot{q}_i\,\dot{q}_j\right)\mathbf{r}_E^{(4)}\right),$$

$$\mathbf{u} = \left(u_j\right)_{j=1,\ldots,4} = \frac{1}{\left(a_{E,a}^{(0)}\right)^2}\mathbf{J}_1\mathbf{T}^{(4)}\mathbf{r}_E^{(4)}\mathbf{C}_1 + \frac{1}{\left(b_{E,a}^{(0)}\right)^2}\mathbf{J}_2\mathbf{T}^{(4)}\mathbf{r}_E^{(4)}\mathbf{C}_2,$$

$$v = \frac{1}{\left(a_{E,a}^{(0)}\right)^2}\left(\left(\mathbf{C}_1\dot{\mathbf{q}}\right)^2 + \mathbf{J}_1\mathbf{T}^{(4)}\mathbf{r}_E^{(4)}d_1\right) +$$

$$+ \frac{1}{\left(b_{E,a}^{(0)}\right)^2}\left(\left(\mathbf{C}_2\dot{\mathbf{q}}\right)^2 + \mathbf{J}_2\mathbf{T}^{(4)}\mathbf{r}_E^{(4)}d_2\right),$$

$a_{E,a}^{(0)}$, $b_{E,a}^{(0)}$ are the elliptical trajectory semi-major and semi-minor axes, respectively.

Vector of dependent velocities are as follows:

$$\dot{\mathbf{q}}_{d_c} = -\mathbf{K}_{d_c}^{-1}\mathbf{K}_{i_c}\dot{\mathbf{q}}_{i_c}, \tag{6}$$

where:

$$\mathbf{K}_{d_c} = \begin{bmatrix}u_2 & u_4\\ c_{12} & c_{14}\end{bmatrix}, \mathbf{K}_{i_c} = \begin{bmatrix}u_1 & u_3\\ c_{11} & c_{13}\end{bmatrix}.$$

The GPME equations supplemented by the equations of programmed constraints take the form:

$$\begin{bmatrix} \mathbf{M}_i\big|_{i\in i_c} + \sum_{j\in i_{d_c}} \mathbf{M}_j \frac{\partial \dot{q}_j}{\partial \dot{q}_i} \\ \mathbf{u} \\ \mathbf{C}_3 \end{bmatrix} \ddot{\mathbf{q}} =$$

$$= \begin{bmatrix} \mathbf{h}_i + \mathbf{Q}_i + \sum_{j\in i_{i_c}\cup i_{d_c}} \dot{q}_j \frac{\partial Q_j}{\partial \dot{q}_i} + \dots \\ \dots \sum_{k\in i_{d_c}} \left(h_k + Q_k + \sum_{j\in i_{i_c}\cup i_{d_c}} \dot{q}_j \frac{\partial Q_j}{\partial \dot{q}_k} \right) \frac{\partial \dot{q}_k}{\partial \dot{q}_i} \\ -v \\ -d_3 + \ddot{z}^{(0)}_{E,a}(t) \end{bmatrix}, \tag{7}$$

where:

$$\mathbf{M} = \sum_{l=1}^{n_l} \mathbf{M}^{(l)},\ \mathbf{M}^{(l)} = \left(m_{ij}^{(l)} \right)_{i,j=1,\dots,n_{dof}^{(l)}},$$

$$m_{ij}^{(l)} = \mathrm{tr}\left\{ \mathbf{T}_i^{(l)} \mathbf{H}^{(l)} \left(\mathbf{T}_j^{(l)} \right)^T \right\},$$

$$\mathbf{h} = \sum_{l=1}^{n_l} \mathbf{h}^{(l)},\ \mathbf{h}^{(l)} = \left(h_i^{(l)} \right)_{i=1,\dots,n_{dof}^{(l)}},$$

$$h_i^{(l)} = \sum_{m=1}^{n_{dof}^{(l)}} \sum_{n=1}^{n_{dof}^{(l)}} \mathrm{tr}\left\{ \mathbf{T}_m^{(l)} \mathbf{H}^{(l)} \left(\mathbf{T}_{m,n}^{(l)} \right)^T \right\} \dot{q}_m^{(l)} \dot{q}_n^{(l)} +$$

$$+ 2\sum_{m=1}^{n_{dof}^{(l)}} \sum_{n=1}^{n_{dof}^{(l)}} \mathrm{tr}\left\{ \mathbf{T}_m^{(l)} \mathbf{H}^{(l)} \left(\mathbf{T}_{i,n}^{(l)} \right)^T \right\} \dot{q}_m^{(l)} \dot{q}_n^{(l)},$$

$$\mathbf{T}_i^{(l)} = \frac{\partial \mathbf{T}^{(l)}}{\partial q_i^{(l)}},\ \mathbf{T}_{i,j}^{(l)} = \frac{\partial^2 \mathbf{T}^{(l)}}{\partial q_i^{(l)} \partial q_j^{(l)}},$$

$$\mathbf{Q} = -(\mathbf{g} + \mathbf{d} + \mathbf{f}) + \mathbf{t},$$

$$\mathbf{g} = \sum_{l=1}^{n_l} \mathbf{g}^{(l)},\ \mathbf{g}^{(l)} = \left(g_i^{(l)} \right)_{i=1,\dots,n_{dof}^{(l)}},\ g_i^{(l)} = m^{(l)} g \mathbf{J}_1 \mathbf{T}_i^{(l)} \mathbf{r}_{C^{(l)}}^{(l)},$$

$$\mathbf{f} = \begin{bmatrix} f^{(1)} & 0 & \mathbf{f}^{(3)} & 0 \end{bmatrix}^T,$$

$$f^{(1)} = s_z^{(sup)} z^{(1)}, \mathbf{f}^{(3)} = \begin{bmatrix} 0 & \mathbf{S}^{(3,1)} \tilde{\mathbf{q}}^{(3,1)} & \dots & \mathbf{S}^{(3,n_{zde}^{(3)})} \tilde{\mathbf{q}}^{(3,n_{zde}^{(3)})} \end{bmatrix},$$

$$\mathbf{d} = \begin{bmatrix} d^{(1)} & 0 & \mathbf{d}^{(3)} & 0 \end{bmatrix}^T,$$

$$d^{(1)} = d_z^{(sup)} z^{(1)}, \mathbf{d}^{(3)} = \begin{bmatrix} 0 & \mathbf{D}^{(3,1)} \dot{\tilde{\mathbf{q}}}^{(3,1)} & \dots & \mathbf{D}^{(3,n_{zde}^{(3)})} \dot{\tilde{\mathbf{q}}}^{(3,n_{zde}^{(3)})} \end{bmatrix},$$

$\mathbf{H}$ - pseudo-inertia matrix, g - the acceleration of gravity.

3 Simulation study – programmed motion of the crane end-effector

Simulation studies are carried out for two cases: the manipulator with rigid links and flexible support and the manipulator with the flexible link and flexible support. The following parameters are used in simulation: $s_z^{(sup)} = 10^4\ \mathrm{Nm^{-1}}$, $d_z^{(sup)} = 2.5\ \mathrm{Nsm^{-1}}$, $a_{E,a}^{(0)} = 0.375\ \mathrm{m}$, $b_{E,a}^{(0)} = 0.750\ \mathrm{m}$. It is assumed that the end-effector $z_E^{(0)}$ coordinate changes according to the following function:

$$z_{E,a}^{(0)}(t) = z_0 + A\cos\left(\frac{2\pi}{T} t \right), \tag{8}$$

where z_0 is the initial position of the end-effector, A is the amplitude and T is a time period of oscillations. It is assumed that $A = 0.5\,\mathrm{m}$ and $T = 2.5\,\mathrm{s}$. The column (2) is driven by the driving torque defined by:

$$t_{dr}^{(1)}(t) = \begin{cases} 10 t_{dr,0}^{(1)} \left(\frac{t}{t_0} \right)^3 - 15 t_{dr,0}^{(1)} \left(\frac{t}{t_0} \right)^4 + 6 t_{dr,0}^{(1)} \left(\frac{t}{t_0} \right)^5, & t < t_0, \\ t_{dr,0}^{(1)}, & t \ge t_0. \end{cases} \tag{9}$$

It is assumed $t_0 = 5\,\mathrm{s}$ and $t_{dr,0}^{(1)} = 1\ \mathrm{Nm}$.

The dynamic equations of motion are integrated using 4th order Runge-Kutta scheme with using the Baumgarte stabilization method for the programmed constraint equations. Figure 2 shows time courses of the joint coordinates obtained from simulations.

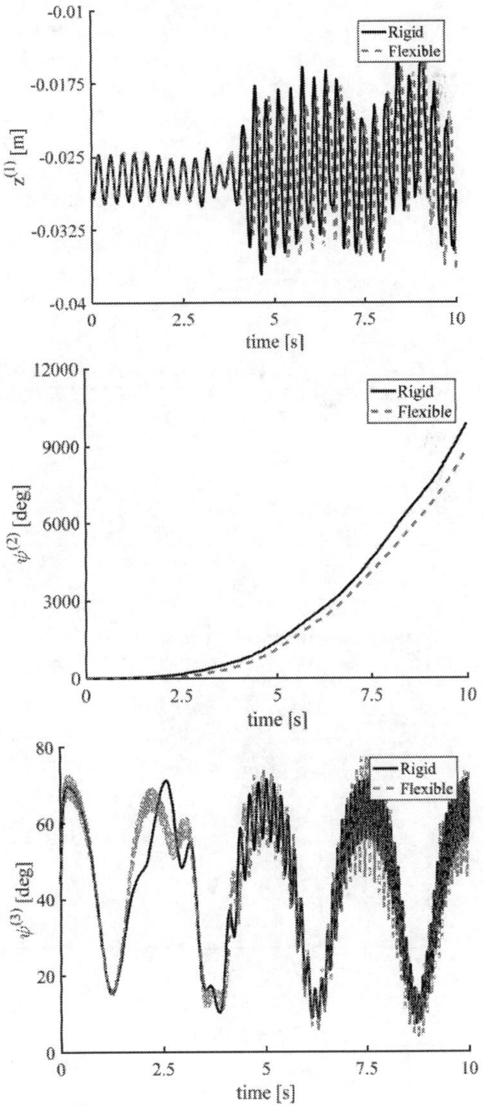

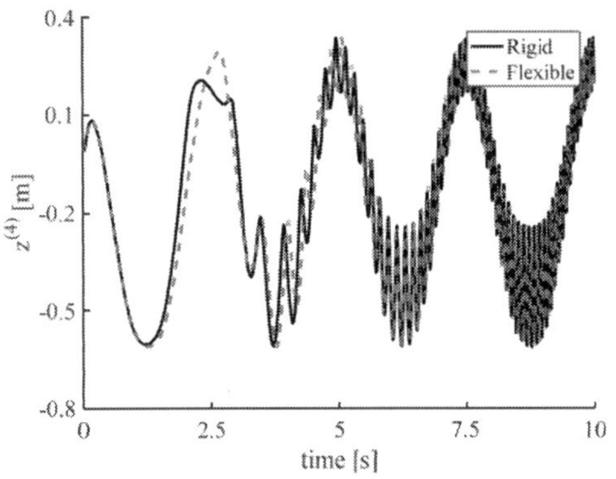

Figure 2. Time course of the joint coordinates

It can be observed that the link flexibility has significant influence on motion of the manipulator. The results show that the programmed motion of the manipulator is compensated by the appropriate combination of the rotation angle of the arm 3 and the displacement of the slider 4. When the flexibility is taken into account additional vibrations with small amplitude and large frequency are induced at initial phase of the motion. After 5 s, the amplitude of vibrations of arm 3 is larger than those obtained for the rigid body model.

Figure 3 presents the course of programmed constraints.

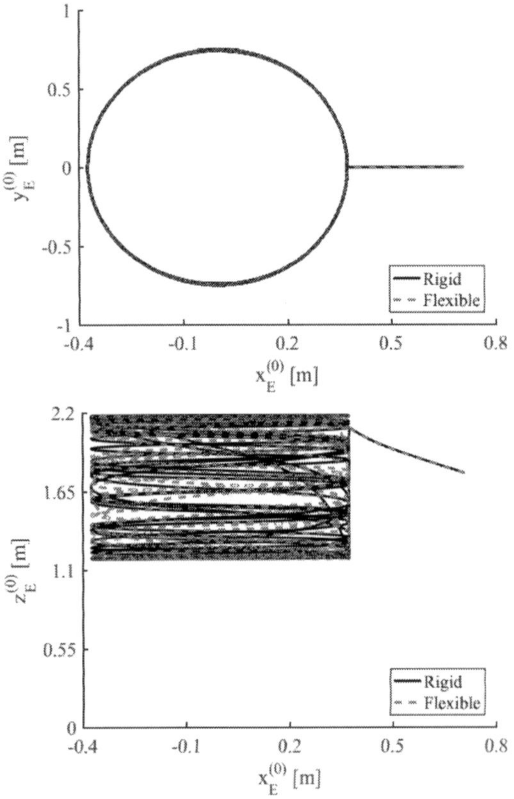

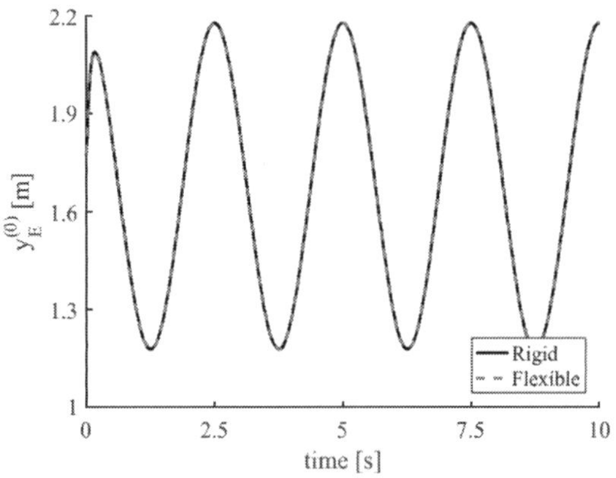

Figure 3. The end-effector trajectory and the time course of $z_E^{(0)}$ coordinate

It can be observed a good agreement of the obtained courses with the assumed programmed constraints. At the initial phase of motion the end-effector is located outside the assumed trajectory and after some period of time this trajectory is achieved.

4 Conclusions

The computational reference model of a crane equipped with the flexible link and support subjected to first order programmed constraints is presented. It is generated by the automatic computational algorithm designed using the formalism of joint coordinates and homogeneous transformations. The algorithm enables constraints incorporation into the system dynamics and modeling flexibility of the system parts and supports. The presented approach can be generalized for systems with higher order constraints.

References

1. A.A. Ata, H. Johar, International Journal of Advanced Robotic Systems, **1**, 2 (2004)
2. A.A. Ataa, W.F. Faresb, M.Y. Sa'adehc, Procedia Engineering, **41** (2012)
3. E. Jarzębowska, K. Augustynek, A. Urbaś, Concurrency and Computation: Practice and Experience (2018), doi:10.1002/cpe.4452. 25
4. E. Jarzębowska, K. Augustynek, A. Urbaś, Proc. of ASME Design Eng. Techn. Conf., No UNSP V006T10A016 (2017)

Numerical Approximation of Lyapunov-Exponents for Quasiperiodic Motions

Robert Fiedler[1,*] and *Hartmut* Hetzler[1,**]

[1]University of Kassel, Institute of Mechanics, Engineering Dynamics, 34125 Kassel, Germany

Abstract. This paper proposes an approach to approximate the Lyapunov-spectrum of quasiperiodic flows on isolated invariant manifolds numerically. Once the invariant manifold has been determined, integrations over the infinite, one dimensional time interval – as calculating the Lyapunov-spectrum for instance – can be transformed into an integral over a finite, p-dimensional domain, where p is the dimension of the manifold. The application of the proposed approach is demonstrated by calculating the Lyapunov-spectrum of periodic and quasiperiodic motions of a forced van-der-Pol equation. The results are compared to results from a classical time integration based method using a continuous Gram-Schmidt orthonormalization.

1 Introduction & Prerequisites

Dynamical systems excited by mechanisms with one single frequency usually exhibit periodic oscillations. As soon as a second mechanism excites the dynamical system simultaneously, oscillations of different physical origin interfere which may lead to quasiperiodic motions, unless synchronization is existent. Due to the infinite period length the description of quasiperiodic motions using standard methods (which are mainly tailored for periodic processes) is usually either not possible or very disadvantageous. The alternative approach proposed in this contribution is based on isolated (normally hyperbolic) invariant manifolds of the flow, which can be determined by solving a system of partial differential equations (invariance equations) [3, 6–8].

Due to the invariance property, solutions on the manifold will never leave it. Moreover, it is assumed that the manifold is isolated, i.e. neighboring solutions are either attracted or repelled. Finally, it is assumed that the flow fills the manifold densely as $t \to \infty$. Since on the other hand the Lyapunov-exponents converge for time approaching infinity (Oseledec's Theorem), the description using an invariant p-dimensional manifold for quasiperiodic motions basically includes all required informations; here, p is the number of internal (base) frequencies. Now, the transition of the infinite time interval to a finite p-dimensional domain is performed by applying the mean-value theorem [6]. A prerequisite for the application of the mean-value theorem is the presence of a parallel flow on the manifold. Since the method described in [3] – which is based on a physical amplitude-phase-parametrization – does not directly provide parallel flows, a transformation from the physical to a parallel parametrization has been introduced. By calculating the corresponding diffeomorphism, the mean-value theorem becomes applicable.

*e-mail: robert.fiedler@uni-kassel.de
**e-mail: hetzler@uni-kassel.de

2 Calculation of the Lyapunov-exponents

The Lyapunov-exponents are an established quantity to assess the stability of arbitrary motions. In general they describe the local exponential growth or decrease rate of perturbations applied to a known solution. Consider an autonomous dynamical system

$$\dot{\mathbf{z}} = \mathbf{f}(\mathbf{z}), \qquad \mathbf{z} \in \mathbb{R}^n \tag{1}$$

which is linearized about a reference solution $\mathbf{z}_0$. This solution has n Lyapunov-exponents of first order $\sigma_1 \geq \sigma_2 \geq \ldots \geq \sigma_n$ which describe the averaged local exponential decay rates near this solution. Together they constitute the Lyapunov-spectrum. If no exponents is positive the corresponding motion is considered as stable. Beyond these exponents of first order one may define $m^{\text{th.}}$ order exponents $\sigma^{(m)}$ $(m = 1, ..., n)$, which describe the growth or decrease of m-dimensional volumes in the state-space near $\mathbf{z}_0$ and are calculated by all possible sums of m different Lyapunov-exponents of first order [1].
In order to evaluate the Lyapunov-exponents over a finite-dimensional invariant manifold, a connection between time parametrization and the invariant manifold describing the quasiperiodic motion is required.

2.1 The mean-value theorem

In [6] a possible way to connect quantities over a time parametrization and an invariant manifold describing quasiperiodic motions is given through

$$\lim_{T\to\infty} \frac{1}{T} \int_t^{t+T} \mathbf{g}(\mathbf{z}(\tau))\, \mathrm{d}\tau = \frac{1}{(2\pi)^p} \int_0^{2\pi} \ldots \int_0^{2\pi} \mathbf{g}(\bar{\mathbf{z}}(\bar{\boldsymbol{\theta}}))\mathrm{d}\bar{\theta}_1 \ldots \mathrm{d}\bar{\theta}_p, \tag{2}$$

where $\mathbf{g}$ is an arbitrary function, $\mathbf{z}$ is a quasiperiodic solution and $\bar{\boldsymbol{\theta}} = [\bar{\theta}_1, ..., \bar{\theta}_p]$ is the parametrization (torus coordinates) of an invariant manifold exhibiting a parallel flow.

In the following variables with a bar $(\bar{.})$ refer to parallel flows. Since the assumption was made, that an invariant manifold is calculated with the method described in [3], the flow on the manifold is not parallel because of the usage of physical coordinates. But since the parallel flow is a canonical form of the physical flow, there has to be a diffeomorphism connecting both parametrization, hence

$$\begin{aligned} &\frac{1}{(2\pi)^p}\int_0^{2\pi}\cdots\int_0^{2\pi}\mathbf{g}(\bar{\mathbf{z}}(\bar{\boldsymbol{\theta}}))\mathrm{d}\bar{\theta}_1\ldots\mathrm{d}\bar{\theta}_p \\ &= \frac{1}{(2\pi)^p}\int_0^{2\pi}\cdots\int_0^{2\pi}\mathbf{g}(\hat{\mathbf{z}}(\boldsymbol{\theta}))\det\left(\frac{\partial\boldsymbol{\phi}(\boldsymbol{\theta})}{\partial\boldsymbol{\theta}}\right)\mathrm{d}\theta_1\ldots\mathrm{d}\theta_p, \end{aligned} \tag{3}$$

where $\boldsymbol{\theta} = [\theta_1, ..., \theta_p]$ is the physical parametrization (torus coordinates) of an invariant manifold and $\boldsymbol{\phi}(\boldsymbol{\theta}) = \bar{\boldsymbol{\theta}}$ is a diffeomorphism. It is important to note, that if an invariant manifold is calculated with the methods described in [7, 8] a parallel flow is given and therefore section 2.2 can be skipped.

2.2 Invariance equation of the diffeomorphism

In [3] equation (1) is first transformed to amplitudes $\mathbf{A}(\boldsymbol{\theta})$ and phases with a subsequent partitioning of the phases in dependent phase angles $\boldsymbol{\psi}(\boldsymbol{\theta})$ and torus coordinates (parametrization) $\boldsymbol{\theta}$ resulting in

$$\dot{\mathbf{z}} = \mathbf{f}(\mathbf{z}), \qquad \mathbf{z} = \begin{bmatrix}\mathbf{A}\\ \boldsymbol{\theta}\\ \boldsymbol{\psi}\end{bmatrix}, \qquad \mathbf{f}(\mathbf{z}) = \begin{bmatrix}\mathbf{h}(\mathbf{z})\\ \boldsymbol{\Omega}(\mathbf{z})\\ \boldsymbol{\Psi}(\mathbf{z})\end{bmatrix}. \tag{4}$$

Using these for solving the invariance equation the quantities $\mathbf{h}(\mathbf{z})$, $\boldsymbol{\Omega}(\mathbf{z})$ and $\boldsymbol{\Psi}(\mathbf{z})$ are known. By assuming a parallel flow, the equations of motion read:

$$\begin{bmatrix}\dot{\bar{\mathbf{A}}}\\ \dot{\bar{\boldsymbol{\theta}}}\\ \dot{\bar{\boldsymbol{\varphi}}}\end{bmatrix} = \begin{bmatrix}\bar{\mathbf{h}}(\bar{\mathbf{z}})\\ \bar{\boldsymbol{\omega}}\\ \bar{\boldsymbol{\Psi}}(\bar{\mathbf{z}})\end{bmatrix}. \tag{5}$$

It is important to note, that by this assumption the time derivation of the parametrization is constant $\bar{\boldsymbol{\omega}}$. It is easily comprehensible by looking at figure 1. The flow on the invariant manifold represents the solution of the dynamical system (1) in time. By choosing an arbitrary point $(\bar{\theta}_1^*, \bar{\theta}_2^*)$ on the invariant manifold, the change of $\bar{\theta}_1$ and $\bar{\theta}_2$ in time has always to be the same because otherwise there would be no parallel flow and therefore has $\dot{\bar{\boldsymbol{\theta}}}$ to be constant.

In section 2.1 is the diffeomorphism introduced as a function describing the deformation to a parallel flow $\bar{\boldsymbol{\theta}} = \boldsymbol{\phi}(\boldsymbol{\theta})$, whereby a substitution in equation (5) is possible. For further considerations it is sufficient to investigate the resulting equations for the parametrization

$$\frac{\mathrm{d}\boldsymbol{\phi}(\boldsymbol{\theta})}{\mathrm{d}t} = \bar{\boldsymbol{\omega}}. \tag{6}$$

Applying the chain rule

$$\frac{\partial\boldsymbol{\phi}(\boldsymbol{\theta})}{\partial\boldsymbol{\theta}}\frac{\mathrm{d}\boldsymbol{\theta}}{\mathrm{d}t} = \bar{\boldsymbol{\omega}} \tag{7}$$

and identifying the time derivation of the parametrization of the physical flow in (4) lead to the invariance equation of the diffeomorphism

$$\frac{\partial\boldsymbol{\phi}(\boldsymbol{\theta})}{\partial\boldsymbol{\theta}}\boldsymbol{\Omega}(\boldsymbol{\theta}) = \bar{\boldsymbol{\omega}}, \tag{8}$$

where $\boldsymbol{\Omega}(\boldsymbol{\theta})$ is known through the calculation of the invariant manifold. Unknown in this equations are the internal frequencies $\bar{\boldsymbol{\omega}}$ and the wanted diffeomorphism $\boldsymbol{\phi}(\boldsymbol{\theta})$. Equation (8) is just as the invariance equation for the manifold 2π-periodic in $\boldsymbol{\theta}$. Since the diffeomorphism is only in the derivatives existent and does not have any restrictions concerning the parametrization, the solution of equation (8) is not unique. Through additional boundary conditions, e.g. $\boldsymbol{\phi}(\mathbf{0}) = \mathbf{0}$ which corresponds to equal origins of the parametrization with $\boldsymbol{\theta}$ and $\bar{\boldsymbol{\theta}}$, the equation gets solvable.

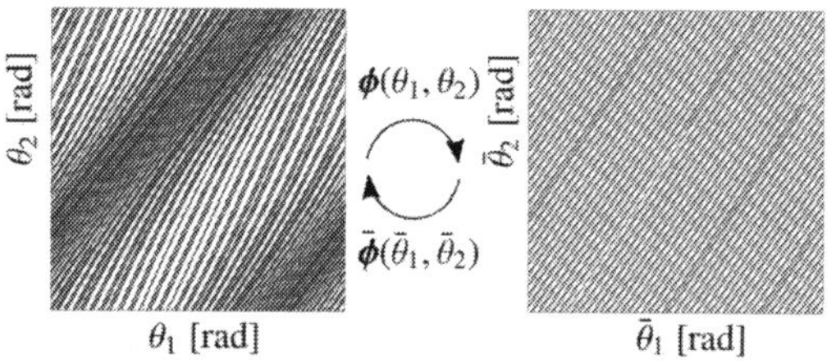

Figure 1. Projection of the flow on a $\mathbb{T}^2$ into the parametrization plane for a finite time of a quasiperiodic solution: (left) physical flow, (right) parallel flow.

2.3 The Lyapunov-exponent of $m^{\text{th.}}$ order

The Lyapunov-exponents of $m^{\text{th.}}$ order describe the characteristic growth or decease of subspace volumes near the considered reference trajectory. It can be shown that these quantities are not unique and depend on the chosen subspace volume – correspondingly different combinations of m exponents are possible. But like the largest Lyapunov-exponent of first order dominates and thus defines the Lyapunov-exponent, the greatest Lyapunov-exponents of $m^{\text{th.}}$ order dominates and defines the Lyapunov-exponent of $m^{\text{th.}}$ order. Considering the Lyapunov-exponent of $n^{\text{th.}}$ order the result is unique, since it defines the sum of all exponents. This leads to a basis-independent formulation, whereby the current direction of each exponent in an orthonormal basis is irrelevant. With Liouville's theorem the change of the volume in time can be deduced

$$\left.\frac{\mathrm{d}V}{\mathrm{d}t}\right|_{t=0} = \int_{V_0}\nabla\cdot\mathbf{f}(\mathbf{z})\,\mathrm{d}V. \tag{9}$$

From this an equation to calculate the Lyapunov-exponent of $n^{\text{th.}}$ order can be derived [1]

$$\sigma^{(n)} = \sum_{i=1}^{n}\sigma_i = \lim_{T\to\infty}\frac{1}{T}\int_t^{t+T}\nabla\cdot\mathbf{f}(\mathbf{z}(\tau))\,\mathrm{d}\tau. \tag{10}$$

Applying the mean-value theorem with respect to the physical flow (3) a criterion for a known manifold results as

$$\sigma^{(n)} = \frac{1}{(2\pi)^p} \int_0^{2\pi} \cdots \int_0^{2\pi} \nabla \cdot \mathbf{f}(\hat{\mathbf{z}}(\boldsymbol{\theta})) \det\left(\frac{\partial \boldsymbol{\phi}(\boldsymbol{\theta})}{\partial \boldsymbol{\theta}}\right) \mathrm{d}\theta_1 \ldots \mathrm{d}\theta_p. \tag{11}$$

This equation can be evaluated once the invariance equation of the diffeomorphism (8) has been solved. Note that this criterion is also applicable on manifolds with a parallel flow, since the Jacobian determinant is then one.

2.4 The LYAPUNOV-spectrum

In contrast to the LYAPUNOV-exponent of $n^{\text{th.}}$ order, the exponents of first order are associated with characteristic directions. This results in numerical difficulties which can be solved by different approaches. An overview can be found in [5]. An established method for the calculation of the whole spectrum or the m largest ones bases upon a differential formulation of the GRAM–SCHMIDT process [4], which is enhanced by a stability parameter in [2]. The equations of motion read

$$\begin{aligned} \dot{\mathbf{z}} &= \mathbf{f}(\mathbf{z}) \\ \dot{\mathbf{e}}_m &= \mathbf{J} \cdot \mathbf{e}_m - \left(G_m \mathbf{e}_m + \sum_{l<m} L_{lm} \mathbf{e}_l\right) \qquad m = 1, ..., k \leq n \\ \dot{\Lambda}_m &= J_{mm}, \end{aligned} \tag{12}$$

where $\mathbf{J}$ is the Jacobian matrix, $\mathbf{E} = [\mathbf{e}_1, ..., \mathbf{e}_k]$ is a set of time-dependent orthonormal vectors and $G_m = J_{mm} + \beta(\mathbf{e}_m \cdot \mathbf{e}_m - 1)$ with the stabilization parameter $\beta > -\sigma_k$, which requires some a priori knowledge. $L_{lm} = J_{lm} + J_{ml} + 2\beta\, \mathbf{e}_l \cdot \mathbf{e}_m$ and $J_{lm} = \mathbf{e}_l \cdot (\mathbf{J} \cdot \mathbf{e}_m)$. The LYAPUNOV-exponents of first order can be calculated by

$$\sigma_m = \lim_{t\to\infty} \frac{1}{t} \Lambda_m \qquad \text{with} \qquad \Lambda_m(0) = 0, \tag{13}$$

whereby they can be seen as the temporal mean value. These values converge almost surly for arbitrary initial values $\mathbf{z}_0$ and $\mathbf{E}_0$ (OSELEDEC's theorem). By integration of $\dot{\Lambda}_m$ under consideration of the initial values and substituting it in equation (13)

$$\sigma_m = \lim_{t\to\infty} \frac{1}{t} \int_0^t \mathbf{e}_m \cdot (\mathbf{J} \cdot \mathbf{e}_m)\, \mathrm{d}\tau \tag{14}$$

results. Considering the mean-value theorem lead to

$$\sigma_m = \frac{1}{(2\pi)^p} \int_0^{2\pi} \cdots \int_0^{2\pi} \mathbf{e}_m \cdot (\mathbf{J} \cdot \mathbf{e}_m) \det\left(\frac{\partial \boldsymbol{\phi}(\boldsymbol{\theta})}{\partial \boldsymbol{\theta}}\right) \mathrm{d}\theta_1 \ldots \mathrm{d}\theta_p, \tag{15}$$

where the basis vectors $\mathbf{e}_m$ are unknown. Since they depend only on time, it is possible to define them on the manifold. Applying the chain rule, a system of partial differential equations

$$\frac{\partial \mathbf{e}_m}{\partial \boldsymbol{\theta}} \boldsymbol{\Omega}(\boldsymbol{\theta}) = \mathbf{J} \cdot \mathbf{e}_m - \left(G_m \mathbf{e}_m + \sum_{l<m} L_{lm} \mathbf{e}_l\right) \tag{16}$$

for calculating the basis vectors results. As long as the spectrum is not degenerated, i.e. when two or more LYAPUNOV-exponents are equal or from a numerical point of view very close to each other, equation (16) has periodic boundary conditions in $\boldsymbol{\theta}$ and has unique $\mathbf{e}_m$. Inserting the results from equations (8) and (16) in (15) a criterion for the LYAPUNOV-spectrum is defined.

3 Application to the forced VAN-DER-POL equation

To demonstrate and analyze the proposed approach, it is applied to the forced VAN-DER-POL equation

$$q'' + \varepsilon\left(q^2 - 1\right) q' + q = \Gamma \cos(\eta\tau), \tag{17}$$

where $\varepsilon = 0.3$ and $\Gamma = 0.6$. The excruciation frequency is chosen as $\eta \in [0.3, 1.5]$. Due to interfering mechanisms of excitation quasiperiodic motions as well as periodic motions (due to synchronization) can be observed. In order to verify the numerical results for the LYAPUNOV-exponents, equation (12) is solved with a time integration for $\tau = [0, 20000]$. Results calculated with a time integration are identified by ${}^{\tau}(.)$.
For a given invariant manifold the calculation of the diffeomorphism to assure for parallel flows is the first task, since both the criterion for $\sigma^{(n)}$ and the one for the LYAPUNOV-spectrum require a parallel flow. If $\sigma^{(n)}$ is of interest, equation (11) can be evaluated with no further calculations. Is the calculation of the spectrum of interest, the invariant basis $\mathbf{E}$ on the manifold has to be determined. Solving equation (16) the basis is known and the criterion described by equation (15) can be evaluated. In figure

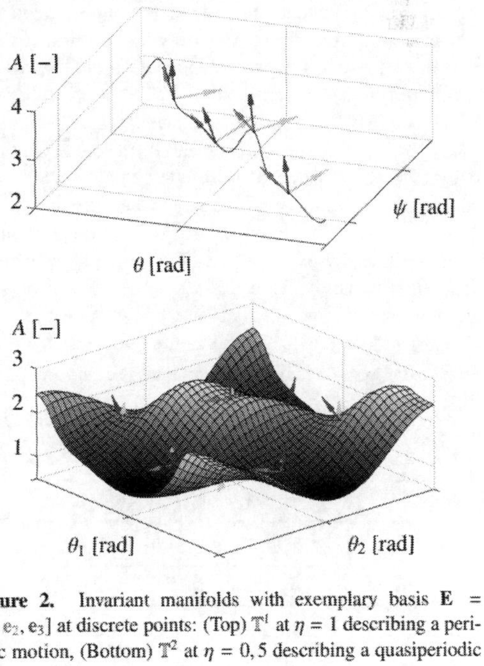

Figure 2. Invariant manifolds with exemplary basis $\mathbf{E} = [\mathbf{e}_1, \mathbf{e}_2, \mathbf{e}_3]$ at discrete points: (Top) $\mathbb{T}^1$ at $\eta = 1$ describing a periodic motion, (Bottom) $\mathbb{T}^2$ at $\eta = 0,5$ describing a quasiperiodic motion (cf. [3])

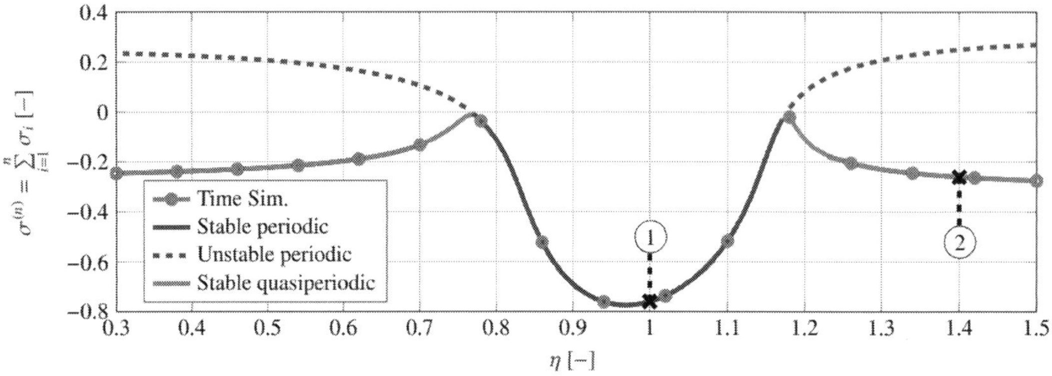

Figure 3. Lyapunov-exponent of $n^{\text{th.}}$ order $\sigma^{(n)}$ plotted against η. Calculated with the proposed criterion for a periodic and quasiperiodic solution and compared to s time simulation using a continuous Gram-Schmidt orthonormalization.

2 are basis at random points depicted. It is noteworthy that the tangential space of the invariant manifolds at the given points is visible and therefore the directions of the Lyapunov-exponents equal to zero. Knowing the invariant basis at every point, the Lyapunov-spectrum can be calculated.

Figure 3 shows the resulting $\sigma^{(n)}$ for invariant manifolds describing periodic and quasiperiodic motions. It is conspicuous that the quasiperiodic solution can not be calculated to the bifurcation point. The reason for that is the parametrization of the underlying method of the manifold approximation. By a modification of the method this behavior can be disabled but should not be discussed here further. The comparison with the time simulation reveals, that both methods exhibit the same results. In the case of the van-der-Pol equation the calculation of the sum is sufficient to state stability of a quasiperiodic motions since a $\mathbb{T}^2$ in $\mathbb{R}^3$ has two Lyapunov-exponents equal to zero. In the case of higher dimensional spaces $\mathbb{R}^n$ with embedded $\mathbb{T}^p$ the necessity of knowing the Lyapunov-spectrum, or at least the largest exponent, is essential for identifying stability of motions, since the absolute value of the sum of the negative exponents can be larger than the one of the positive ones. The outcome is a negative sum containing positive Lyapunov-exponents resulting in an unstable motion. In contrast to that states a positive sum always an unstable motion since then at least one exponent is positive. To demonstrate the identification of the spectrum, the proposed approach is used to calculate it at (1) and (2).

Table 1. Calculated Lyapunov-spectrum. Proposed approach compared to a time simulation

	i	1	2	3
(1)	σ_i	0	−0.1186	−0.6424
(η = 1)	${}^{\tau}\sigma_i$	0	−0.1186	−0.6424
(2)	σ_i	0	0	−0.2614
(η = 1.4)	${}^{\tau}\sigma_i$	0	0	−0.2598

Comparing both methods shows that the results are almost equal. Only σ_3 differs from the time simulation: however, this difference stems from accumulated numerical errors of the underling methods. The advantages resulting from the proposed approach is on the one hand that especially for quasiperiodicity the Lyapunov-spectrum is known without concerns about convergence and on the other hand can the approach be applied to unstable solution which is not possible with time simulations. Reversing the time is not expedient since both the unstable and stable exponents revers their behavior resulting in an unstable motion. Concluding it has to be mentioned that for large systems the calculation of the spectrum is costly, because for equation (16) n^2 equations have to be solved. However the evaluation of equation (11) itself is fast once the invariant manifold has been determined with the methods proposed in [3, 7, 8], for instance.

References

[1] Argyris, J. H., Faust, G., Haase, M. and Friedrich, R., *An Exploration of Dynamical Systems and Chaos: Completely Revised and Enlarged Second Edition* (Springer, 2015), p. 224.

[2] Christiansen, F. and Rugh, H. H., Nonlinearity **10**, 1063-1072 (1997)

[3] Fiedler, R. and Hetzler, H., Proceedings of the 9th European Nonlinear Dynamics Conference, ID 279 (2017)

[4] Goldhirsch, I., Sulem, P. L., and Orszag, S. A., Physica D: Nonlinear Phenomena **27**(3), 311-337 (1987).

[5] Ramasubramanian, K. and Sriram, M. S., Physica D: Nonlinear Phenomena **139**(1-2), 72-86 (2000).

[6] Samoilenko, A. M. and Petryshyn, R., *Multifrequency oscillations of nonlinear systems* (Springer Science & Business Media, Dordrecht, 2004), p. 266.

[7] Schilder, F., Osinga, H. M. and Vogt, W., SIAM Journal on Applied Dynamical Systems **43**, 459-488 (2005).

[8] Schilder, F., Vogt, W., Schreiber, S., and Osinga, H. M., International Journal for Numerical Methods in Engineering **67**, 629-671 (2006).

Improving energy harvesting in excited Duffing harvester device using a delayed piezoelectric coupling

Z. Ghouli[1], M. Hamdi[2], M. Belhaq[1]

[1]Faculty of Sciences Ain Chock, University Hassan II-Casablanca, Morocco
[2]FST-Al Hoceima, Mohamed First University Oujda, Al Hoceima, Morocco

Abstract. The present work examines the influence of time delay introduced in the piezoelectric circuit of an excited Duffing harvester device with hardening stiffness on the vibration and voltage amplitudes. Specifically, we seek to exploit a delayed electrical circuit of the harvester to enhance its performance. We consider the case of a monostable system and we use a perturbation technique to approximate the periodic response and the corresponding voltage amplitude near the principal resonance. It is shown that for appropriate values of delay amplitude, the energy harvesting performance is improved over a certain range of coupling parameters and excitation frequencies. Numerical simulation is conducted to support the analytical predictions.

1 Introduction

A monostable piezoelectric nonlinear energy harvester (EH) device consisting of a Duffing-type harvester with hardening stiffness coupled to a delayed piezoelectric circuit is considered in this paper. The objective is to examines the influence of time delay in the piezoelectric circuit on the amplitude of the response and the voltage amplitude.

Recently, a delayed Duffing-type monostable harvester subject to a harmonic excitation and coupled to a piezoelectric circuit was studied in the case where the time delay is introduced only in the mechanical subsystem [1]. The optimal performance of the harvester device in term of time delay parameters was investigated and it was shown that the induced large-amplitude quasiperiodic vibrations can be used to extract energy over broadband of excitation frequencies away from the resonance.

Here we examine the influence of time delay on EH in the case where the time delay is introduced in the electric circuit. To show the contribution of introducing time delay in the electrical circuit in impoving EH performance, a comparison is carried out with the case where time delay is absent [2].

In the next section we present the harvester system and we derive approximation of the periodic response and the voltage amplitude using the multiple scales method [3]. In Section 3 the influence of different system parameters of the harvester device on the EH performance is analyzed in the presence and absence of time delay. A summary of the results is given in the concluding section.

2 Harvester system, vibration response and voltage amplitude

Consider an excited Duffing oscillator coupled to an electrical circuit through a delayed piezoelectric device as shown in the schematic presented in Fig. 1.

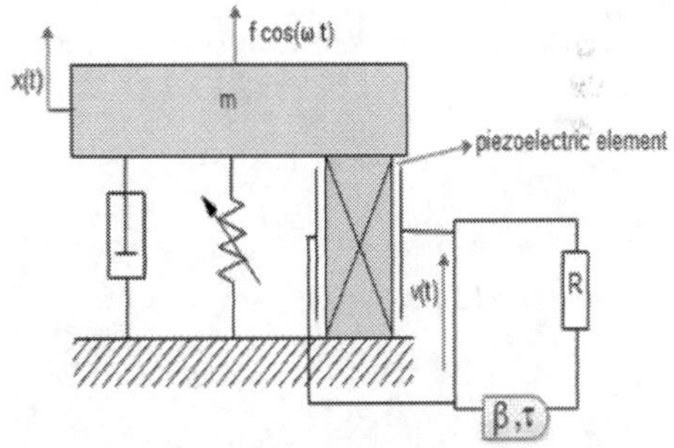

Fig. 1. Schematic description of the EH system

The governing equations for this harvester can be written in the dimensionless form as

$$\ddot{x}(t) + 2\mu\omega_n\dot{x}(t) + \omega_n^2 x(t) + \alpha x(t)^3 - \chi v(t) = f\cos(\omega t) \quad (1)$$

$$\dot{v}(t) + \lambda v(t) + \kappa\dot{x}(t) = \beta v(t-\tau) \quad (2)$$

where $x(t)$ is the relative displacement of the rigid mass m, $v(t)$ is the voltage across the load resistance, ω_n is the natural frequency of the system, μ is the mechanical damping ratio, α is the stiffness parameter, χ is the piezoelectric coupling term in the mechanical attachment, κ is the piezoelectric coupling term in the electrical circuit, λ is the reciprocal of the time constant of the electrical circuit, β and τ are, respectively, the feedback gain and time delay in the electrical circuit, and f, ω are, respectively, the amplitude and the frequency of the excitation. To investigate the influence of time delay in the electric circuit on the performance of the harvester, we approximate the response of the system near the primary resonance by introducing the resonance condition $\omega = \omega_n + \sigma$ where σ is a detuning parameter.

The method of multiple scales [3] is implemented by introducing a bookkeeping parameter ϵ and scalling parameters as $\mu = \epsilon\tilde{\mu}, \alpha = \epsilon\tilde{\alpha}, \chi = \epsilon\tilde{\chi}, \sigma = \epsilon\tilde{\sigma}, f = \epsilon\tilde{f}$.

Thus, Eqs. (1) and (2) take the form

$$\ddot{x}(t) + \omega_n^2 x(t) = \epsilon[-2\tilde{\mu}\omega_n\dot{x}(t) - \tilde{\alpha}x(t)^3 + \tilde{\chi}v(t) + \tilde{f}\cos(\omega t)] \quad (3)$$

$$\dot{v}(t) + \lambda v(t) + \kappa\dot{x}(t) = \beta v(t-\tau) \quad (4)$$

A solution to Eqs. (3) and (4) is given by

$$x(t) = x_0(T_0, T_1) + \epsilon x_1(T_0, T_1) + O(\epsilon^2) \quad (5)$$

$$v(t) = v_0(T_0, T_1) + \epsilon v_1(T_0, T_1) + O(\epsilon^2) \quad (6)$$

where $T_0 = t$, and $T_1 = \epsilon t$. In terms of the variables T_i, the time derivatives become $\frac{d}{dt} = D_0 + \epsilon D_1 + O(\epsilon^2)$ and $\frac{d^2}{dt^2} = D_0^2 + \epsilon^2 D_1^2 + 2\epsilon D_0 D_1 + O(\epsilon^2)$ where $D_i^j = \frac{\partial^j}{\partial^j T_i}$. Substituting (5) and (6) into (3) and (4) and equating coefficient of like powers of ϵ, we obtain up to the second order the systems

$$D_0^2 x_0 + \omega_n^2 x_0 = 0 \quad (7)$$

$$D_0 v_0 + \lambda v_0 + \kappa D_0 x_0 = \beta v_{0\tau} \quad (8)$$

$$D_0^2 x_1 + \omega_n^2 x_1 = -2D_0 D_1 x_0 - 2\tilde{\mu}\omega_n D_0 x_0 - \tilde{\alpha}x_0^3 + \tilde{\chi}v_0 + \tilde{f}\cos(\omega_n T_0 + \tilde{\sigma}T_1) \quad (9)$$

$$D_0 v_1 + \lambda v_1 = -D_1 v_0 - \kappa D_0 x_1 - \kappa D_1 x_0 + \beta v_{1\tau} \quad (10)$$

Up to the first order the solution reads

$$x_0(T_0, T_1) = A(T_1)e^{i\omega_n T_0} + \bar{A}(T_1)e^{-i\omega_n T_0} \quad (11)$$

$$v_0(T_0, T_1) = \frac{-\kappa i\omega_n A(T_1)}{\lambda + i\omega_n - \beta e^{-i\omega_n\tau}}e^{i\omega_n T_0} + \frac{\kappa i\omega_n \bar{A}(T_1)}{\lambda - i\omega_n - \beta e^{i\omega_n\tau}}e^{-i\omega_n T_0} \quad (12)$$

where $A(T_1)$ and $\bar{A}(T_1)$ are unknown complex conjugate functions. Substituting Eqs. (11) and (12) into (9) and (10) and eliminating the secular terms, one obtains

$$-2i\omega_n(D_1 A) - 2i\tilde{\mu}\omega_n^2 A - 3\tilde{\alpha}A^2\bar{A} - \frac{\kappa i\omega_n\tilde{\chi}A}{\lambda + i\omega_n - \beta e^{-i\omega_n\tau}} + \frac{\tilde{f}}{2}e^{i\tilde{\sigma}T_1} = 0 \quad (13)$$

Expressing $A = \frac{1}{2}ae^{i\theta}$ where a and θ are the amplitude and the phase of the modulation, we obtain up to the first order the modulation equations

$$\begin{cases} \dfrac{da}{dt} = S_1 a + S_4\sin(\gamma) \\ a\dfrac{d\gamma}{dt} = S_2 a + S_3 a^3 + S_4\cos(\gamma) \end{cases} \quad (14)$$

where $S_i (i = 1, ..., 4)$ are given in Appendix and $\gamma = \tilde{\sigma}T_1 - \theta$. The steady-state response of system (14), corresponding to periodic oscillations of Eqs. (3) and (4), are determined by setting $\frac{da}{dt} = \frac{d\gamma}{dt} = 0$. Eliminating the phase, we obtain the following sixth-order algebraic equation in a

$$(S_1 a)^2 + (S_2 a + S_3 a^3)^2 = S_4^2 \quad (15)$$

which can b written in the form

$$Ba^6 + Ca^4 + Da^2 + E = 0 \quad (16)$$

where $B = S_3^2$, $C = 2S_2S_3$, $D = S_1^2 + S_2^2$ and $E = -S_4^2$. The discriminant of Eq. (16) reads

$$\Delta = \frac{P^3}{27} + \frac{Q^2}{4} \quad (17)$$

where $P = \frac{D}{B} - \frac{C^2}{3B^2}$ and $Q = \frac{2}{27}(\frac{C}{B})^3 - \frac{CD}{3B^2} + \frac{E}{B}$. Equation (16) has three real positive roots if Δ is negative and one positive root if Δ is positive.

The solution given by Eqs. (11) and (12) can be written as $x_0(T_0, T_1) = a\cos(\omega_n t + \theta)$ and $v_0(T_0, T_1) = V\cos(\omega_n t + \theta + \arctan\frac{\lambda - \beta\cos(\omega_n\tau)}{\omega_n + \beta\sin(\omega_n\tau)})$ such that the condition $\omega_n + \beta\sin(\omega_n\tau) \neq 0$ is satisfied. Moreover, the voltage amplitude V is given by

$$V = \frac{\kappa\omega_n}{\sqrt{(\lambda - \beta\cos(\omega_n\tau))^2 + (\omega_n + \beta\sin(\omega_n\tau))^2}}a \quad (18)$$

3 Main results

Next, the influence of different parameters of the system on vibration and voltage amplitudes is examined using Eqs. (15) and (18), respectively.

Figures 2a and 2b show, respectively, the variation of the amplitude of the periodic vibrations and the voltage V versus σ for the excitation amplitude $f = 0.1$. The undelayed case ($\beta = 0$) is presented by a grey line and the delayed one ($\beta \neq 0$) is indicated by a black line. The analytical prediction (solid lines) is compared to results obtained by numerical simulation using `dde23` algorithm [4] for the delayed case and the method of Runge Kutta of order 4 for the undelayed case. It can be observed from the figures that the EH performance is improved by introducing the time delay in the electric circuit, especially near the resonance peak.

Figure 3 shows the variation of the amplitude of the periodic response and the voltage amplitude V versus σ for larger value of the forcing amplitude $f = 1.5$. The grey line corresponds to the undelayed case ($\beta = 0$) and the black one is obtained in the delayed one. As in the previous case (Fig. 2 for $f = 0.1$) the voltage amplitude is increased around the resonance peak in the case of delayed circuit.

Similarly, the variation of the amplitude of vibration and voltage versus the piezoelectric coupling term in the electrical circuit κ are depicted in Fig. 4 showing the range of parameter κ where a substantial EH performance is achieved (Fig. 4b). The variation of the amplitudes of vibration and voltage as a function of the piezoelectric coupling term in the mechanical attachment χ are shown in Fig. 5 for $\beta = 0$ (undelayed circuit, grey line) and $\beta = \lambda$ (delayed circuit, black line). Inspection of Fig. 5 shows that for values of σ taken relatively away from the resonance, the range of χ where a better performance of EH is obtained (Fig. 5b).

4 Conclusion

We have studied the EH performance in a monostable Duffing forced oscillator with hardening stiffness coupled to a delayed piezoelectric harvester device. The analysis was carried out near the primary resonance and a perturbation method was performed to obtain approximation of vibration and voltage amplitudes. The influence of time delay introduced in the piezoelectric subsystem on the EH performance of the harvester was studied. Results showed that

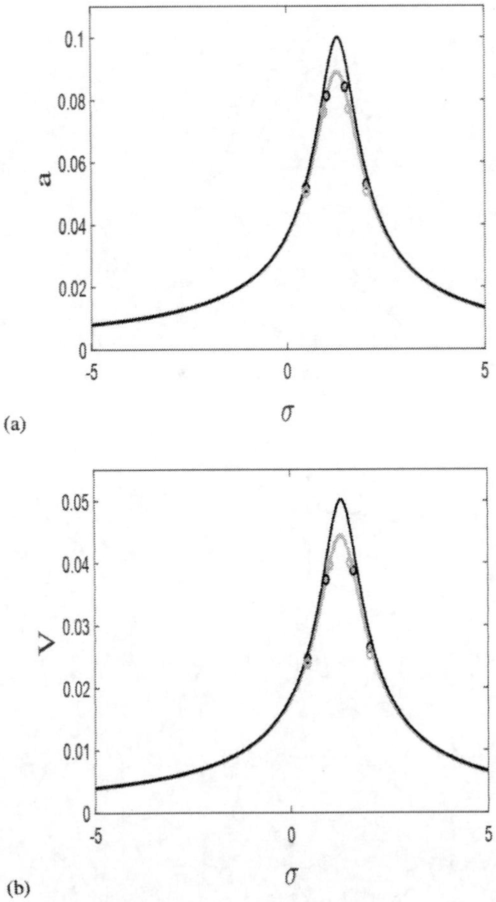

Fig. 2. Vibration (a) and voltage (b) amplitudes vs σ for $f = 0.1, \chi = 0.05, \kappa = 0.5, \omega_n = 1, \mu = 0.5, \alpha = 5, \lambda = 0.05$, and $\tau = 6.2$. Black (grey) line for delayed (undelayed) electric circuit, $\beta = \lambda$ ($\beta = 0$). Analytical prediction (solid lines); numerical simulation (black (grey) circles for delayed (undelayed) electric circuit).

in the presence of time delay in the electrical circuit the voltage amplitude increases at certain range of coupling parameters especially in the vicinity of the resonance peak. In other words, by tuning the delay amplitude in the electric circuit, it is possible to optimize the power output performance over certain range of the piezoelectric coupling terms.

5 Appendix

$$S_1 = -\mu\omega_n - \frac{\chi\kappa(\lambda - \beta\cos(\omega_n\tau))}{2[(\lambda - \beta\cos(\omega_n\tau))^2 + (\omega_n + \beta\sin(\omega_n\tau))^2]}$$

$$S_2 = \sigma - \frac{\chi\kappa(\omega_n + \beta\sin(\omega_n\tau))}{2[(\lambda - \beta\cos(\omega_n\tau))^2 + (\omega_n + \beta\sin(\omega_n\tau))^2]}$$

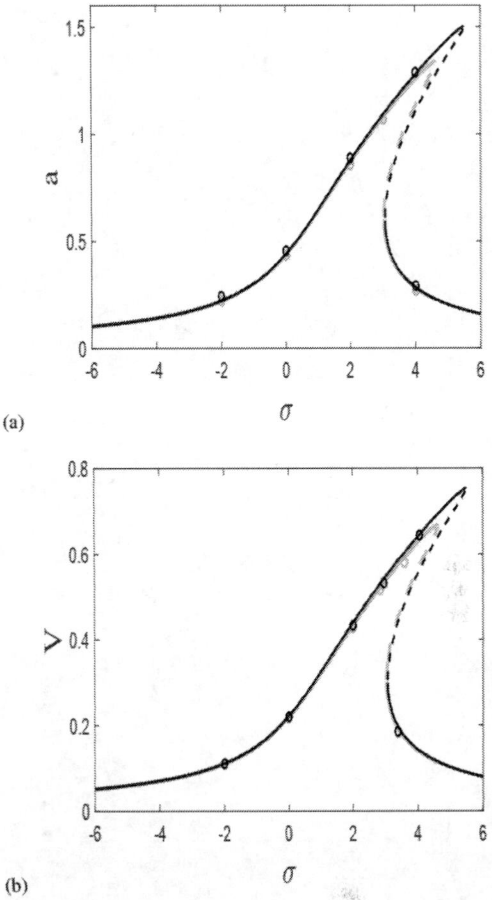

Fig. 3. Vibration (a) and voltage (b) amplitudes vs σ for $f = 1.5, \chi = 0.05, \kappa = 0.5, \omega_n = 1, \mu = 0.5, \alpha = 5, \lambda = 0.05$, and $\tau = 6.2$. Black (grey) line for delayed (undelayed) electric circuit $\beta = \lambda$ ($\beta = 0$). Analytical prediction (solid lines for stable and dashed line for unstable) and numerical simulation (Black (grey) circles for delayed (undelayed) electric circuit).

$$S_3 = -\frac{3\alpha}{8\omega_n}$$

$$S_4 = \frac{f}{2\omega_n}$$

References

1. Z. Ghouli, M. Hamdi, F. Lakrad, M. Belhaq, Quasiperiodic energy harvesting in a forced and delayed Duffing harvester device, J. Sound Vib. **407** (2017) 271-285.
2. A. Erturk, D.J. Inman, Piezoelectric Energy Harvesting, John Wiley, 2011.
3. Nayfeh, A.H., Mook, D.T.: Nonlinear Oscillations. Wiley, New York, 1979.

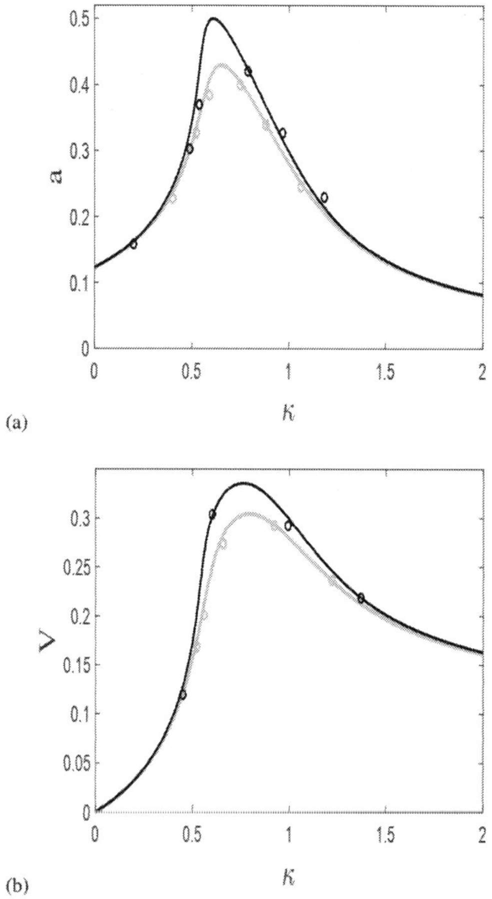

Fig. 4. Vibration (a) and voltage (b) amplitudes vs κ for $\sigma = 2, \chi = 5, f = 0.5, \omega_n = 1, \mu = 0.5, \alpha = 5, \lambda = 0.05$, and $\tau = 6.2$. Black (grey) line for delayed (undelayed) electric circuit $\beta = \lambda$ ($\beta = 0$). Analytical prediction (solid lines) and numerical simulation (Black (grey) circles for delayed (undelayed) electric circuit).

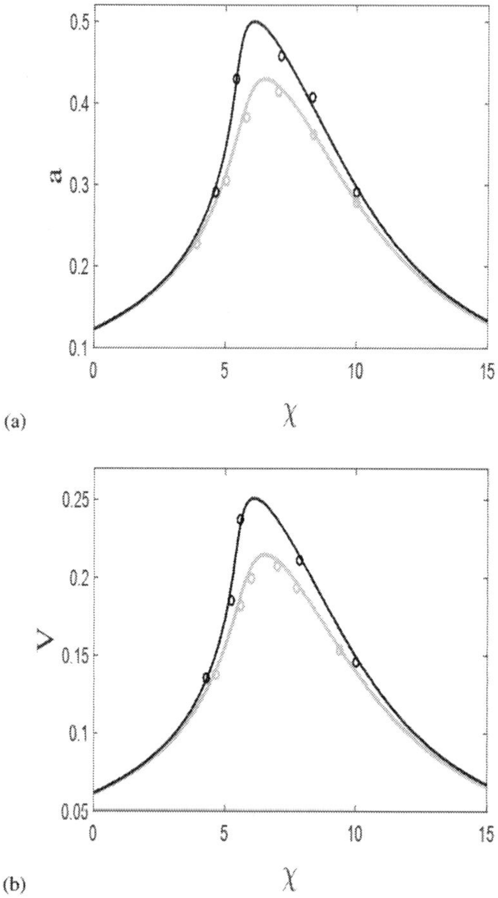

Fig. 5. Vibration (a) and voltage (b) amplitudes vs χ for $\sigma = 2, \kappa = 0.5, f = 0.5, \omega_n = 1, \mu = 0.5, \alpha = 5, \lambda = 0.05$, and $\tau = 6.2$. Black (grey) line for delayed (undelayed) electric circuit $\beta = \lambda$ ($\beta = 0$). Analytical prediction (solid lines) and numerical simulation (Black (grey) circles for delayed (undelayed) electric circuit).

4. L.F. Shampine, S. Thompson, Solving delay differential equations with dde23. PDF available on-line at http: //www.radford.edu/~ thompson/webddes/tutorial.pdf (2000).

Multi-speed Multi-load Bearing Diagnostics Using Extended Phase Space Topology

T. Haj Mohamad[1,*], C. A. Kitio Kwuimy[2,**], and C. Nataraj[1,***]

[1]Villanova Center for Analytics of Dynamic Systems (VCADS), Villanova University, Villanova, PA 19085, USA
[2]Department of Engineering Education, College of Engineering and Applied Science, University of Cincinnati, Cincinnati, OH 45221, USA

Abstract. This paper presents the application of Extended Phase Space Topology (EPST) and conventional statistical time domain features in the diagnostics of various bearing faults in rotating machinery systems. Bearings with various health statuses operating under multiple motor loads and speeds are analyzed. The results indicate remarkable performance in detecting anomalous behavior and in identifying faults with virtually perfect accuracy, recall and precision.

1 Introduction

There is considerable interest in rolling element bearing diagnostics due to their wide use in machines, and their failure is one of the most common reasons for machines breaking down. Predicting the potential breakdown of a machine can prevent catastrophic failures and production loss. Accurate prediction can also improve safety and reduce repair costs in most industries.

To this end, rolling element bearings are traditionally diagnosed using envelope analysis techniques [1–3]. Although these techniques are capable of extracting detailed information from the response of the system, they rely on expert knowledge and have to be carefully designed and adapted to a specific problem. For many applications, such as online/automated diagnostics, it is desirable to develop robust algorithms that can provide accurate results in order to identify faults and their severity quickly and in an automated fashion.

Our previous work [4–6] introduced and developed a family of methods based on the phase space characterization for different dynamical systems. In the present paper, we apply a novel feature extraction technique that we call Extended Phase Space Topology (EPST) [7–9] in order to detect and identify bearings with different health statuses under multiple motor operating conditions of load and speed.

The rest of this paper is organized as follows: In section 2, the experimental setup and measurement of data are introduced. Section 3 describes the mathematical details of the proposed feature extraction method. Section 4 discusses the results of fault identification. Finally, section 5 summarizes and concludes this paper.

*e-mail: thajmoha@villanova.edu
**e-mail: kwuimyck@ucmail.uc.edu
***e-mail: c.nataraj@villanova.edu

2 Experimental Setup

The proposed approach in this study was implemented on data collected from motor bearings provided by Case Western Reserve University [10, 11]. The experimental setup that was used in the study is shown in Fig. 1. The bearing experimental setup (shown in Fig. 1) consists of a 2-horsepower (hp) motor with a torque transducer/encoder and a dynamometer. The dynamometer was used to apply various loads on the motor. For this study, four motor loads of 0 to 3 hp (motor speeds of 1797 to 1720 RPM) were considered. Moreover, SKF deep groove ball bearings were used for the drive end (DE) and fan end (FE) bearings. Multiple test DE bearings were used with different health conditions including one healthy bearing (H) and three defective bearings with a ball defect (B), an inner race defect (IR) and an outer race defect (OR). Table 1 summarizes the collected data under various operating conditions.

Bearings with defective ball and inner race were seeded with single point faults with diameters of 0.007, 0.014 and 0.021 inches while bearings with outer race defect were seeded with faults of 0.007 and 0.014 diameter. For outer race defective bearings, the faults were seeded at a 6 o'clock position. The vibrational signals were recorded using accelerometers installed on the drive end and fan end bearing housing at a 12 o'clock position. The signal was recorded for approximately 20 seconds and 10 seconds for healthy and defective bearing conditions, respectively, at a sampling rate of 12,000 Hz.

3 Methodology

An overview of the fault detection procedure used in this paper is summarized in Fig. 2 and the procedure is described as follows. First, acceleration data of the drive end and fan end bearings were collected for various bearing

Table 1. Bearing data description of various conditions, fault diameters, motor loads and speeds.

Bearing Condition	Fault Diameter	Motor Load	Motor Speed
Healthy	0	0	1797
		1	1772
		2	1750
		3	1730
Ball/ Inner race/ Outer race	0.007/ 0.014	0	1797
		1	1772
		2	1750
		3	1730
Ball/ Inner race	0.021	0	1797
		1	1772
		2	1750
		3	1730

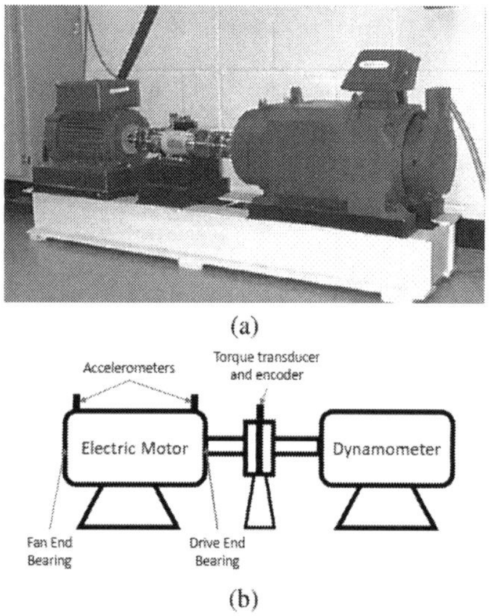

Figure 1. (a): The bearing experimental test [10]. (b): Schematic diagram of the test rig.

conditions, motor loads and shaft rotational speeds. The vibration data was then divided into multiple segments of one-second periods. For each segment, two sets of features were extracted using EPST and conventional statistical time domain features. The EPST method is based on characterizing the topology of the density distribution of the vibration data. The density distribution of the vibration signal is approximated using Legendre polynomials. The coefficients of the orthogonal polynomials are then used as features. Our previous work [7–9] explains in detail the derivation and the implementation of the EPST method. A linear support vector machine (SVM) with optimal parameters was built in order identify the bearing condition i.e., H, B, IR and OR using features of the EPST method.

In order to extract the EPST features, let $X=(x_1, x_2, ..., x_n)$ be an independent and identically distributed sample data drawn from a distribution with an unknown density function Ψ. The shape of this function can be estimated by its kernel density estimator (^ indicates that it is an estimate, and h indicates that its value can depend on h).

$$\hat{\Psi}_h(x) = \frac{1}{nh}\sum_{i=1}^{n}\Gamma\left(\frac{x-x_i}{h}\right) \quad (1)$$

where, $h > 0$ is a smoothing parameter called the bandwidth, and $\Gamma(.)$ is the kernel function which satisfies the following requirements.

$$\int_{-\infty}^{\infty}\Gamma(u)\,du = 1 \quad (2)$$

$$\Gamma(-u) = \Gamma(u)\ \forall u \quad (3)$$

There is a range of kernel functions that can be used, including uniform, triangular, biweight, triweight, Epanechnikov, normal, etc. Due to its conventional and convenient mathematical properties, we use the standard normal density function in our approach, defined as the following:

$$\Gamma(u) = \frac{1}{\sqrt{2\pi}}e^{-\frac{1}{2}u^2} \quad (4)$$

Let x be a state of the system and $y_d = \hat{f}_h(x)$, its density computed using the kernel density estimator. y_d is then approximated with Legendre orthogonal polynomials. Legendre polynomials can be directly obtained from Rodrigues' formula which is given by:

$$\Phi_m(x) = \frac{1}{2^m m!}\frac{d^m}{dx^m}\left[(x^2-1)^m\right], \quad m = 0, 1, 2, \ldots \quad (5)$$

It can also be obtained using Bonnet's recursion formula:

$$(m+1)\Phi_{m+1}(x) = (2m+1)x\Phi_m(x) - m\Phi_{m-1}(x) \quad (6)$$

where the first two terms are given by:

$$\Phi_0(x) = 1, \quad \Phi_1(x) = x \quad (7)$$

The coefficients of the Legendre polynomials are obtained by using the least squares method assuming the following linear regression model:

$$\Psi(x,\beta) = \sum_{j=1}^{m}\beta_j\Phi_j(x) \quad (8)$$

Letting

$$X_{ij} = \frac{\partial \Psi(x_i, \beta)}{\partial \beta_j} = \Phi_j(x_i), \quad (9)$$

the estimated coefficients are given by:

$$\hat{\beta} = (X^T X)^{-1} X^T y_d \quad (10)$$

The coefficients $\hat{\beta}$ constitute the features in our approach that can be used in classification or regression problems. The approximated density using Legendre Polynomials is then calculated using the following:

$$\Psi_a = X\hat{\beta} \quad (11)$$

Root mean square error (RMSE) and Pearson's correlation coefficient (PCC) were calculated to compute the quality of the fit using the following equations:

$$\text{RMSE} = \sqrt{\frac{1}{N} Z Z^T}, \quad \text{PCC} = \frac{\sigma_d^T \sigma_a}{\sqrt{(\sigma_d^T \sigma_d)(\sigma_a^T \sigma_a)}} \quad (12)$$

where, $Z = (y_d - \Psi_a)$ is the residual vector, N is the number of points in the density function, $\sigma_d = (y_d - E\{y_d\})$ and $\sigma_a = (\Psi_a - E\{\Psi_a\})$. $E\{.\}$ is the expected value.

In addition to the EPST features, various statistical features were extracted from the vibration data including both simple and high order statistics. Eight statistical features were computed including (1) mean, (2) root mean square (RMS), (3) standard deviation (SD), (4) peak value (PV), (5) minimum value (MV), (6) crest factor (CF), (7) kurtosis value (KV), and (8) skewness value (SV).

4 Results and Discussion

The EPST method, which is based on characterizing the topology of the density distribution of the vibration data, is applied by mapping the data onto a density space for each condition and then approximating these densities using Legendre polynomials. After the density plots were estimated, the plots were then approximated using Legendre polynomials of order 30. The order of Legendre polynomials was selected based on the best fit between the actual and the approximate densities by calculating the root mean square error (RMSE) and Pearsons Correlation Coefficient (PCC).

The extracted features were used as input to a linear SVM classifier in order to classify various bearing conditions (H, B, IR and OR). For each set of data, 36 features were used in the feature set for training (10 EPST features and 8 statistical features for each acceleration signal). Various SVMs were trained using 60% of the data samples (276 cases) and a cross validation algorithm with five folds. 40% of the data samples (184 cases) were used for testing the classifier. The box constrained parameter was optimized by minimizing the k-fold cross validation loss in order to find the optimal linear SVM.

To describe the performance of the classifier for the fault detection application, various metric rates were calculated from the confusion matrix such as precision, recall and overall accuracy. For each class, precision measures the rate of correct predictions out of all predictions that were made by the classifier. In other words, when the classifier predicts a class, precision indicates how often the prediction is correct. Recall measures the correctly predicted rate of the actual samples for a given class. If the classifier has high recall and low precision for a certain class, this means that the classifier is biased to that class. A high precision and low recall classifier for a given class indicates that the classifier is too conservative. The classifier is less likely to predict the given class, but when it does predict it, it is very likely to be correct. Ideally, a classifier with high recall and high precision is what we seek. Finally, the overall accuracy of the classier is the rate of the correct prediction.

The results can be summarized as follows.

- The classifier model was able to identify a hyperplane in the feature space that can separate all training set samples with 100% overall accuracy as shown in Table 2.
- The classification of the test data presented by means of the confusion matrix in Table 3 indicate an overall accuracy of 100% for predicting different bearing conditions.
- The classifier was able to detect anomaly behavior in the system by distinguishing between healthy and defective bearing conditions.
- The classifier was able to predict all the cases with 100% accuracy, 100% recall and 100% precision for each bearing condition.
- The extracted features provide valuable information about the bearing defect regardless of the fault level and severity.
- No a priori knowledge of the system was included in the extracted features. This implies the EPST approach can be conveniently applied to diverse dynamical systems in an automated process with minimal need for adaptation or reliance on expert knowledge about the system.

5 Conclusion

The study demonstrates that the EPST method combined with statistical measures provide very accurate references about the status of the health of bearings by exploiting the dynamical response. An optimal SVM was achieved using the extracted features in order to predict different bearing conditions. Furthermore, results show that the innovative EPST procedure has outstanding performance in fault detection and identification with a virtually perfect classification accuracy. For future work, a comparison with other bearing diagnostics techniques is necessary. Also, a severity analysis is needed to predict the fault diameter and the defect level, and is being pursued.

References

[1] R.B. Randall, J. Antoni, Mechanical systems and signal processing **25**, 485 (2011)

[2] V. Patel, N. Tandon, R. Pandey, Measurement **45**, 960 (2012)

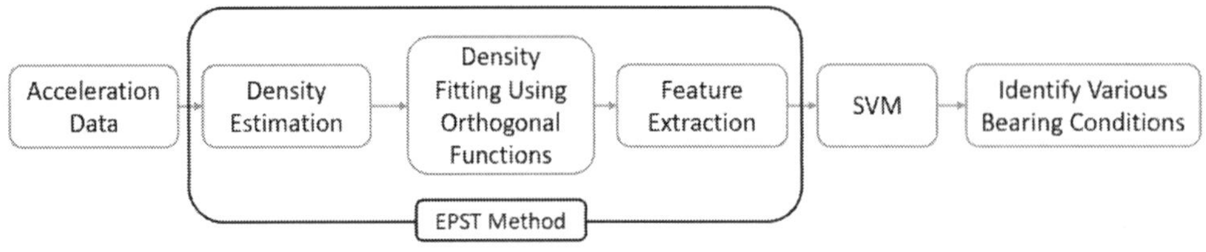

Figure 2. An overview of the bearing diagnostics method

Table 2. Prediction of various defect classes for training data.

Output Class	Training Confusion Matrix: H	B	IR	OR	Recall	Precision
H	84 30.4%	0 0.0%	0 0.0%	0 0.0%	100.0%	100.0%
B	0 0.0%	72 26.1%	0 0.0%	0 0.0%	100.0%	100.0%
IR	0 0.0%	0 0.0%	72 26.1%	0 0.0%	100.0%	100.0%
OR	0 0.0%	0 0.0%	0 0.0%	48 17.4%	100.0%	100.0%
	H	B	IR	OR	**Accuracy 100.0%**	
	Target Class					

Table 3. Prediction of various defect classes for test data

Output Class	Test Confusion Matrix: H	B	IR	OR	Recall	Precision
H	56 30.4%	0 0.0%	0 0.0%	0 0.0%	100.0%	100.0%
B	0 0.0%	48 26.1%	0 0.0%	0 0.0%	100.0%	100.0%
IR	0 0.0%	0 0.0%	48 26.1%	0 0.0%	100.0%	100.0%
OR	0 0.0%	0 0.0%	0 0.0%	32 17.4%	100.0%	100.0%
	H	B	IR	OR	**Accuracy 100.0%**	
	Target Class					

[3] Y. Ming, J. Chen, G. Dong, Mechanical Systems and Signal Processing **25**, 1773 (2011)

[4] M. Samadani, C.K. Kwuimy, C. Nataraj, *Diagnostics of a nonlinear pendulum using computational intelligence*, in *ASME 2013 Dynamic Systems and Control Conference* (American Society of Mechanical Engineers, 2013)

[5] M. Samadani, C.K. Kwuimy, C. Nataraj, Communications in Nonlinear Science and Numerical Simulation **20**, 583 (2015)

[6] T. Haj Mohamad, C. Nataraj, *Gear Fault Diagnostics Using Extended Phase Space Topology*, in *Annual Conference of the Prognostics and Health Management Society* (2017)

[7] T. Haj Mohamad, M. Samadani, C. Nataraj, Journal of Vibration and Acoustics (2018)

[8] M. Samadani, T. Haj Mohamad, C. Nataraj, *Feature Extraction for Bearing Diagnostics Based on the Characterization of Orbit Plots With Orthogonal Functions*, in *ASME 2016 International Design Engineering Technical Conferences and Computers and Information in Engineering Conference* (American Society of Mechanical Engineers, 2016), pp. V008T10A026–V008T10A026

[9] T. Haj Mohamad, C.A.K. Kwuimy, C. Nataraj, *Discrimination of Multiple Faults in Bearings Using Density-Based Orthogonal Functions of the Time Response*, in *ASME 2017 International Design Engineering Technical Conferences and Computers and Information in Engineering Conference* (American Society of Mechanical Engineers, 2017), pp. V008T12A035–V008T12A035

[10] K. Loparo, http://www. eecs. cwru. edu/laboratory/bearing/download. htm (2003)

[11] W.A. Smith, R.B. Randall, Mechanical Systems and Signal Processing **64**, 100 (2015)

The dynamics of HIV infection model with logistic growth and infected cells in eclipse phase

Sanaa Harroudi[a] and Karam Allali[b]

Department of Mathematics, FSTM, Laboratory of Mathematics & Applications , University Hassan II-Casablanca, PO Box146, Mohammedia, Morocco

Abstract. In this paper, we study a mathematical model of human immunodeficiency virus dynamics with logistic growth and infected cells in eclipse phase. This model describes the interactions between uninfected CD4+ T cells, infected CD4+ T cells in latent stage, productively infected CD4+ T cells and free virus. The positivity and boundedness of solutions for non negative initial data are proved. The stability of disease-free equilibrium and endemic equilibrium are rigorously established. Numerical simulations are also provided to give a more complete representation of the system dynamics.

Keywords: HIV infection, Logistic growth, Stability, Viral dynamics.

1 Introduction

Human immunodeficiency virus (HIV) is a pathogen which causes the well known acquired immunodeficiency syndrome (AIDS). HIV continues to be a major global public health issue, having claimed more than 35 million lives so far; in 2016 more than 1 million people died from HIV-related causes globally [1]. Without treatment of the HIV infection, HIV advances in stages getting worse over time [2]. However, the most powerful antiretrovirals can not completely eliminate the virus because it remains dormant in some cells [3]. In the last decades, many mathematical models have proved to be useful for describing and understanding the dynamics of HIV infection by considering three classes of $CD4^+$ T cells: uninfected cells, infected cells in latent phase and productively infected cells [4–6]. In recent years, much works extend the basic model by including a logistic growth term that describes the growth rate of healthy $CD4^+$ T cells [7,8], since the proliferation rate of T-cells is density-dependent with the rate of proliferation slowing as the T-cell count gets high [7]. In this study, we extend the recent work [9] by considering a logistic growth rate instead of constant one and the model that we consider is the following

$$\begin{cases} \dot{x} = rx(t)(1 - \dfrac{T(t)}{T_m}) - d_1x(t) - \dfrac{k_1x(t)v(t)}{x(t)+v(t)}, \\ \dot{y} = \dfrac{k_1x(t)v(t)}{x(t)+v(t)} - (d_2+k_2)y(t), \\ \dot{s} = k_2y(t) - d_3s(t), \\ \dot{v} = as(t) - d_4v(t), \end{cases} \quad (1)$$

With

$$T(t) = x(t) + y(t) + s(t).$$

The initial data are

$$x(0) = x_0 \geq 0, y(0) = y_0 \geq 0, s(0) = s_0 \geq 0, v(0) = v_0 \geq 0,$$

$$T_m \geq T_0 = x_0 + y_0 + s_0 > 0.$$

In this model, $x(t)$, $y(t)$, $s(t)$, $v(t)$ represent the concentration of uninfected $CD4^+$ T cells, infected $CD4^+$ T cells in latent stage, productively infected $CD4^+$ T cells and free virus (HIV), respectively. The uninfected $CD4^+$ T cells grow at a rate r, die at a rate d_1 and become infected by free virus a rate k_1. T_m is the carrying capacity of the T-cell population. Infected $CD4^+$ T cells in latent stage are produced at a rate k_1, die at a rate d_2 and become productively infected cells at a rate k_2. Productively infected $CD4^+$ T cells are produced at a rate k_2 and die at a rate d_3. Free virus (HIV) is produced from infected cells at a rate a and die a rate d_4.

The paper is organized as follows. The next section is devoted to the analysis of the model which contain the existence, positivity, boundedness of solutions and the stability analysis. Section 3, deals with some numerical simulations of the model. Finally, we conclude in the last section.

2 Analysis of the model

2.1 Steady states

There exist two steady states of the studied model: The infection-free equilibrium $E_f = (\frac{T_m(r-d_1)}{r}, 0, 0, 0)$ which represent the disease free equilibrium and correspond to the maximal level of healthy $CD4^+$ T cells. $E^* = (x^*, y^*, s^*, v^*)$ is a state of persistent, chronic HIV infection. Explicitly, E^* requires

$$x^* = \frac{T_md_3d_4(d_2+k_2)(R^* - R_0)}{r(d_4(d_3+k_2)(R_0-1)+ak_2)},$$

[a] e-mail: sanaa.harroudi@gmail.com
[b] e-mail: allali@hotmail.com

$$y^* = \frac{T_m d_3^2 d_4^2 (d_2 + k_2)(R_0 - 1)(R^* - R_0)}{ark_2(d_4(d_3 + k_2)(R_0 - 1) + ak_2)},$$

$$s^* = \frac{T_m d_3 d_4^2 (d_2 + k_2)(R_0 - 1)(R^* - R_0)}{ar(d_4(d_3 + k_2)(R_0 - 1) + ak_2)},$$

$$v^* = \frac{T_m d_3 d_4 (d_2 + k_2)(R_0 - 1)(R^* - R_0)}{r(d_4(d_3 + k_2)(R_0 - 1) + ak_2)},$$

where

$$R^* = \frac{(r - d_1)ak_2 + d_3 d_4 (d_2 + k_2)}{d_3 d_4 (d_2 + k_2)}.$$

2.2 Positivity and boundedness of solutions

For the problem deal with cell population evolution, the cell densities should remain non-negative and bounded. In this subsection, we will establish the positivity and boundedness of solutions of the model (1). First of all, for biological reasons, the parameters x_0, y_0, s_0 and v_0 must be larger than or equal to 0. Hence, we have the following result

Proposition 21 *The solutions of the problem* (1) *exist. Moreover, they are bounded and nonnegative for all* $t > 0$.

Proof Notice that system (1) is locally lipschitzian at $t = 0$. Hence the solution of this system exists and is unique on $[0, b)$ for some $b > 0$. Observe that if $x(0) = 0$, then $x \equiv 0$ for all $t > 0$. Thus, we assume below that $x(0) > 0$. We also have the following:

$\dot{y}\mid_{y=0}= \dfrac{k_1 x v}{x+v} \geq 0$, $\dot{s}\mid_{s=0}= k_2 y \geq 0$ and $\dot{v}\mid_{v=0}= as \geq 0$.

This shows that $x(0) = x_0 > 0, y(0) = y_0 \geq 0, s(0) = s_0 \geq 0$ and $v(0) = v_0 \geq 0$ for all $t \in [0, b)$.

On the other hand, for the boundedness of the solutions, we have the following:

$$\frac{dT(t)}{dt} = rx(t)(1 - \frac{T(t)}{T_m}) - d_1 x(t) - d_2 y(t) - d_3 s(t),$$

since

$$T(t) \leq T_m \text{ and } x(t) \leq T(t),$$

we obtain

$$\frac{dT(t)}{dt} \leq rT(t)(1 - \frac{T(t)}{T_m}) \leq rT(t),$$

thus

$$T(t) \leq T_0 e^{-rt},$$

with $T_0 = x_0 + y_0 + s_0$.

We conclude that T is bounded, which means also that x, y and s are bounded.

From the last equation of (1), we have

$$v(t) \leq v(0)e^{-d_4 t} + a \int_0^t s(t)e^{(\xi - t)d_4} d\xi,$$

therefore

$$v(t) \leq v(0) + \frac{a}{d_4} \|s\|_\infty (1 - e^{-d_4 t}).$$

Since $(1 - e^{-d_4 t}) \leq 1$, we conclude that v is bounded.

2.3 The stability analysis

First, the jacobian matrix of the system (1) is given by

$$J = \begin{pmatrix} r(1 - \frac{2x+y+s}{T_m}) - d_1 - \frac{k_1 v^2}{(x+v)^2} & -\frac{rx}{T_m} & -\frac{rx}{T_m} & -\frac{k_1 x^2}{(x+v)^2} \\ \frac{k_1 v^2}{(x+v)^2} & -(d_2 + k_2) & 0 & \frac{k_1 x^2}{(x+v)^2} \\ 0 & k_2 & -d_3 & 0 \\ 0 & 0 & a & -d_4 \end{pmatrix} \quad (2)$$

A straightforward calculation gives the following expression for the basic reproductive number in the model (1):

$$R_0 = \frac{ak_1 k_2}{d_3 d_4 (d_2 + k_2)}.$$

2.3.1 Stability of the infection-free equilibrium point E_f

Here, we will analyze locally asymptotical stability and globally asymptotical stability of the disease-free equilibrium E_f.

Proposition 22 *The free equilibrium point* E_f *is locally asymptotically stable when* $R_0 < 1$ *and unstable when* $R_0 > 1$.

Proof The Jacobian matrix at E_f is given by

$$J_{E_f} = \begin{pmatrix} -(r - d_1) & -(r - d_1) & -(r - d_1) & -k_1 \\ 0 & -(d_2 + k_2) & 0 & k_1 \\ 0 & k_2 & -d_3 & 0 \\ 0 & 0 & a & -d_4 \end{pmatrix}$$

The characteristic polynomial of J_{E_f} is

$$P_{E_f}(\xi) = (\xi + (r - d_1))[\xi^3 + a_1 \xi^2 + a_2 \xi + a_3],$$

where

$$a_1 = d_2 + d_3 + d_4 + k_2,$$
$$a_2 = d_3 d_4 + (d_3 + d_4)(d_2 + k_2),$$
$$a_3 = d_3 d_4 (d_2 + k_2) - ak_1 k_2.$$

While $\xi_1 = -(r - d_1)$ is a negative eigenvalue, the other three eigenvalues are given by the solution of the following cubic equation,

$$\xi^3 + a_1 \xi^2 + a_2 \xi + a_3 = 0.$$

It is clear that, $a_1 > 0$, $a_2 > 0$ and $a_1 a_2 - a_3 = d_3 d_4 (d_3 + d_4) + (d_2 + k_2)(d_3 + d_4)(d_2 + d_3 + d_4 k_2) + ak_1 k_2 > 0$. If $R_0 < 1$, then $a_3 > 0$. From the Routh-Hurwitz Theorem given in [**?**], all roots of this equation have negative real parts. Then E_f is locally asymptotically stable when $R_0 < 1$.

Proposition 23 *The free equilibrium point* E_f *is globally asymptotically stable when* $R_0 < 1$, *provided the initial data satisfies.*

Proof From positivity of solutions, we have that s and v satisfy the differential inequality

$$\frac{ds}{dt} \leq k_2 y(t) - d_3 s(t),$$

$$\frac{dv}{dt} \leq a s(t) - d_4 v(t).$$

It is clear from (1) that if $(s(t), v(t)) \to (0,0)$ then $x(t) \to T_m$ and $y(t) \to 0$. We can apply Theorem 3.2 of [10] and conclude that $(s(t), v(t)) \to (0,0)$ as $t \to \infty$ when $R_0 < 1$. Thus, E_f is globally asymptotically stable.

2.3.2 Stability of the endemic equilibrium point E^*

In this part, we discuss the local stability of the endemic infection equilibrium point E^*.

Proposition 24 *If $R^* < R_0$ or $R_0 < 1$, the point E^* does not exist. The endemic equilibrium point E^* is locally asymptotically stable when $R_0 > 1$ and $R^* > R_0$.*

Proof From the expression of E^* we observe that this point exists when $1 \leq R_0 \leq R^*$ and it becomes E_f when $R_0 = 1$. We assume that $1 \leq R_0 \leq R^*$.
The Jacobian matrix at the endemic equilibrium point E^* is given by

$$J_{E^*} = \begin{pmatrix} r(1 - \frac{2x^*+y^*+s^*}{T_m}) - d_1 - \frac{k_1 v^{*2}}{(x^*+v^*)^2} & -\frac{rx^*}{T_m} & -\frac{rx^*}{T_m} & -\frac{k_1 x^{*2}}{(x^*+v^*)^2} \\ \frac{k_1 v^{*2}}{(x^*+v^*)^2} & -(d_2+k_2) & 0 & \frac{k_1 x^{*2}}{(x^*+v^*)^2} \\ 0 & k_2 & -d_3 & 0 \\ 0 & 0 & a & -d_4 \end{pmatrix} \quad (3)$$

The characteristic polynomial of J_{E^*} is

$$P_{E^*}(\xi) = \xi^4 + b_1\xi^3 + b_2\xi^2 + b_3\xi + b_4,$$

From the Routh-Hurwitz theorem applied to the fourth order polynomial, the eigenvalues of the jacobian matrix (3) have negative real parts since we have $b_1 b_2 > b_3$ and $b_1 b_2 b_3 > b_3^2 + b_1^2 b_4$. Consequently, we obtain the asymptotic local stability of the endemic point E^*.

3 Numerical simulations

In order to carried out the numerical simulations, we have used the Euler finite-difference scheme method. The parameter values or ranges used are presented in Table 1.

Figure 1, shows the behavior of the infection during the first 800 days of observation. We clearly see that the curves converge to the disease-free steady state $E_f = (805, 0, 0, 0)$. In this case the basic reproduction number is less than unity ($R_0 = 3.597 \times 10^{-3} < 1$) which supports our theoretical result about the stability of E_f. Figure 2, shows the evolution of the infection during the first 800 days of observation. We use a same set of parameter values as those in Table 1, but we vary the value of r. For $r = 1$ all the curves converge to the endemic equilibrium $E^* = (443.237, 152.462, 290.353, 1675.439)$, $R_0 = 4.78 > 1$ and $R^* = 10.427$ which prove the stability of this endemic equilibrium. For $r = 2$ all the curves also converge to the endemic equilibrium $E^* =$

Table 1. Parameters, their symbols and default values used in HIV literature.

Parameters	Value	References
r	0.03–3 day^{-1}	[7]
T_m	1500 mm^{-3}	[7]
d_1	0.0139 $days^{-1}$	[9]
k_1	2.5×10^{-4}–0.5 $mm^{-3} day^{-1}$	[9]
d_2	0.0495 $days^{-1}$	[9]
k_2	1.1 $days^{-1}$	[9]
d_3	0.5776 $days^{-1}$	[9]
a	2–1250 $virion\ cell^{-1}$	[9]
d_4	0.3466–2.4 day^{-1}	[9]

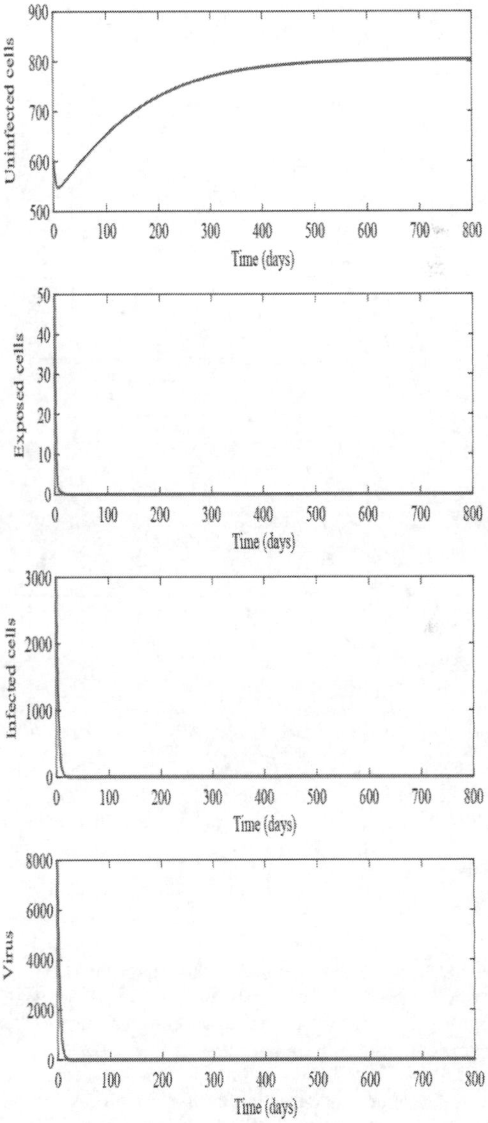

Fig. 1. Behavior of the infection dynamics for $r = 0.03$, $T_m = 1500$, $k_1 = 0.0027$, $d_1 = 0.0139$, $d_2 = 0.0495$, $k_2 = 1.1$, $d_3 = 0.5776$, $a = 5$ and $d_4 = 2.4$, which correspond to the stability of the free-equilibrium point E_f.

 https://doi.org/10.1051/matecconf/201824101012

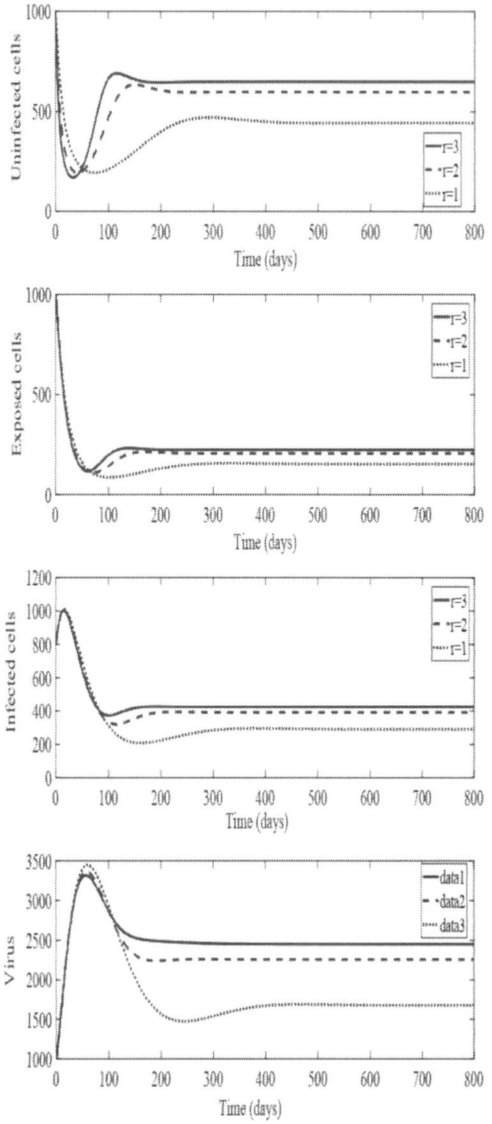

Fig. 2. Behavior of the infection dynamics for $T_m = 1500$, $k_1 = 0.5$, $d_1 = 0.0139$, $d_2 = 0.0495$, $k_2 = 1.1$, $d_3 = 0.5776$, $a = 2$ and $d_4 = 0.3466$ which correspond to the stability of the endemic-equilibrium point.

$(596.797, 205.282, 390.946, 2255.895)$ which prove the stability of this endemic equilibrium, here $R_0 = 4.78 > 1$ and $R^* = 19.987$. For $r = 3$ all the curves converge also to the endemic equilibrium $E^* = (647.984, 222.889, 424.477, 2449.380)$, $R_0 = 4.78 > 1$ and $R^* = 29.547$ which prove the stability of this endemic equilibrium. For all cases $R_0 > 1$ and $R^* > R_0$, the numerical simulations agree with our theoretical finding concerning the stability of the free and endemic equilibria. With low value of r, a significant reduce of infected cells in latent stage, productively infected cells and the viral lowed is observed. One can conclude that the logistic growth of the uninfected cells plays an essential role in the HIV viral dynamics.

4 Conclusion

In this work, we have studied a mathematical model of HIV dynamics with logistic function that describes the interactions between HIV and $CD4^+$ T cells. The considered model includes four differential equations describing the interaction between the uninfected $CD4^+$ T cells, infected $CD4^+$ T cells in latent stage, productively infected $CD4^+$ T cells and free virus (HIV). First, the existence, positivity and boundedness of solutions are proved. Next, both the local stability and global stability of the disease-free equilibrium and local stability of the endemic equilibria are established. Some numerical simulations are performed in order to confirm the theoretical results concerning the equilibrium stability.

References

1. Wordld Health Organization HIV/AIDS facts, http://www.who.int/mediacentre/factsheets/fs360/en/index.html,accessed July (2017).
2. M.A. Chesney, J. ICKOVICS, F.M. HECHT, G. Sikipa, J. Rabkin, Adherence: a necessity for successful HIV combination therapy, AIDS (London, England). **13** (1999) 271-8.
3. S. Pankavich, The Effects of Latent Infection on the Dynamics of HIV, Differential Equations and Dynamical Systems. **24** (2016) 281-303.
4. L. Rong, M.A. Gilchrist, Z. Feng, A.S. Perelson, Modeling within-host HIV1 dynamics and the evolution of drug resistance: tradeoffs between viral enzyme function and drug susceptibility, Journal of Theoretical Biology. **247** (2007) 804-818.
5. L. Rong, A.S. Perelson, Asymmetric division of activated latently infected cells may explain the decay kinetics of the HIV-1 latent reservoir and intermittent viral blips, Mathematical biociences. **217** (2009) 77-87.
6. B. Buonomo, C. Vargas-De-Leon, Global stability for an HIV-1 infection model including an eclipse stage of infected cells, Journal of Mathematical Analysis and Applications. **285** (2012) 709-720.
7. Y. Wang, Y. Zhou, J. Wu, J. Heffernan, Oscillatory viral dyanamics in a delayed HIV pathogenesis model, Mathematical biociences. **209** (2009) 104-112.
8. L. Wang, M.Y. Li, Mathematical analysis of the global dynamics of a model for HIV infection of $CD4^+$ T cells, Mathematical biociences. **200** (2006) 44-57.
9. Q. Sun, L. Min, Y. Kuang, Global stability of infection-free state and endemic infection state of a modified human immunodeficiency virus infection model, IST systems biology. **9** (2015) 95-103.
10. S. A. Gourley, Y. Kuang and J. D. Nagy, Dynamics of a delay differential equation model of hepatitis B virus infection, Journal of Biological Dynamics. **2** (2008) 140-153.

Optimization of vibration energy localization in quasi-periodic structures

Mariem Hbaieb[1,2,*], *Najib* Kacem[1], *Mohamed Amine* Ben Souf[2], *Noureddine* Bouhaddi[1], *Mohamed* Haddar[2]

[1]Univ. Bourgogne Franche-Comté, FEMTO-ST Institute, CNRS/UFC/ENSMM/UTBM, Department of Applied Mechanics, 25000 Besançon, France
[2]Research Laboratory of Mechanics Modeling and Production, National Engineering School of Sfax, Univ. Sfax, 3038 Sfax, Tunisia

Abstract. A mechanical periodic structure in presence of component perturbations can be a seat of a localization of vibration energy. In fact, it is well known that mistuned components have larger response levels than those of perfect components. This results in a localized energy, which can be tapped via harvesting devices. In this study, the dynamic behavior of a quasi-periodic system consisting in weakly connected linear oscillators is investigated. The main objective is to optimize the mistuning parameter, the coupling stiffness and the damping coefficient in order to functionalize the imperfection, which leads to the maximization of the localized vibration energy.

1 Introduction

The presence of irregularities in nominally periodic structures can inhibit the propagation of vibrations within the structure [1]. These small perturbations, defined as mistuning, generally increase the maximum amplitude of the forced response in a well-defined region of the quasi-periodic structure.

Depending on the magnitude of disorder and on the strength of internal coupling for the system, the irregularities may localize the vibration modes and confine the vibrational energy to a region close to the source. [1].

Modern structures are built slender, multi-span and many a times same basic substructure is repeated, for example turbine blades, multi span bridges and repetitive space structures [2]. In fact, in a real situation, there will always be a slight difference between the substructures that are supposed to be perfectly identical because of the geometry, properties of the materials, manufacturing tolerances or degradation in service.

As a result, mistuning may cause high vibration stresses and can have detrimental impact on the high cycle fatigue life of the component. [3].

In order to avoid the phenomena of localization of vibratory energy unexpected, several studies are oriented towards the optimization of the geometry of the nominally periodic structure for the reduction of the effects of disorder [4, 5].

Another approach seeks to amplify the localized vibratory energy in order to store it and use it in several applications such as energy harvesting [6, 7], energy trapping, vibration isolation…

In this paper, an oscillator system, represented for coupled cyclic structures, is chosen to study the combined effects of mismatch, stiffness coupling, and damping on the vibrational localization of cyclic structures.

In this study, the dynamic behavior of a quasi-periodic system consisting in weakly connected linear oscillators is investigated. The studied system involves eight oscillators coupled via linear springs while including two perturbations on two degrees of freedom (DOFs). The simplicity of the system makes it possible to effectively understand the characteristics of the vibration localization phenomenon involved in the cyclic structures.

The main objective of the present study is to optimize the parameters of the quasi-periodic structure in order to maximize the localized vibration energy.

The multi-objective optimization procedure consists in the maximization of the two mistuned DOFs response amplitudes using NSGA algorithm and Pareto front. Several numerical simulations have been performed to validate the proposed approach and to highlight the Pareto-optimal solutions, which can be used as a decision-making tool for the design of quasi-periodic structures with well mastered energy trapping.

2 System Modeling

Figure 1 shows the model for a N-oscillators system subjected to horizontal base excitation. Oscillators are connected to each other by spring of stiffness k_c and each mass is grounded by springs of stiffness k. c is assumed to be the proportional mechanical damping constant and m_n is the mass of each oscillator.

* Corresponding author: mariem.hbaieb@enis.tn

Mass of first oscillator is taken as nominal mass. Mistuning is introduced by varying the mass of two oscillators.

The development of equation of motion of a coupled oscillators is presented in this section.

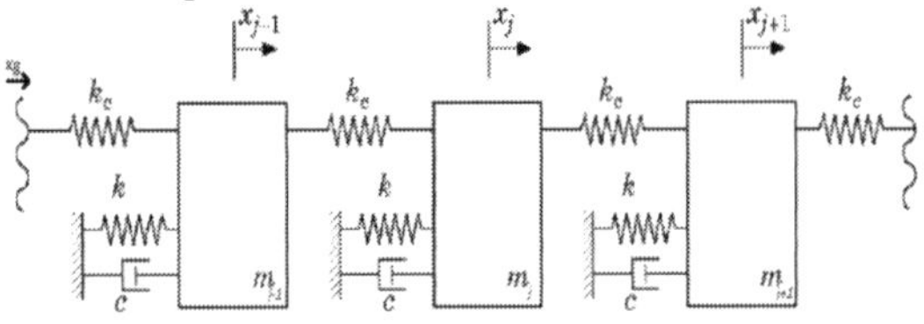

Fig. 1. A n-DOF spring-mass system

Using the Lagrangien approach, the equations of motion can be written as follows.

$$\begin{cases} m_n\ddot{x}_n + c\dot{x}_n + (k+2k_c)x_n - k_c x_{n-1} - k_c x_{n+1} = -m_n\ddot{x}_g \\ \quad with \;\; n = 1,..,N \end{cases} \tag{1}$$

with represent a system of coupled linear equations, subjected to a basic excitation. We suppose that

$$\omega_0 = \sqrt{\frac{k}{m}}, \quad c = 2am\omega_0 \quad , \alpha_n = \frac{m_n}{m}, \; \beta = \frac{k_c}{k} \tag{2}$$

$$x_n = X_n e^{i\omega t} \quad and \quad x_g = X_g e^{i\omega t} \tag{3}$$

where ω_0 is the normal frequency of the sub-structure, α_n is the mistuning parameter, β is the coupling stiffness and a is the damping coefficient.

Then, we obtain the following equations:

$$\begin{cases} \left[-\alpha_n\omega^2 + 2\alpha_n a\omega_0\omega j + \omega_0^2(1+2\beta)\right]X_n - \beta\omega_0^2 X_{n-1} - \beta\omega_0^2 X_{n+1} = \alpha_n\omega^2 X_g \\ \quad with \;\; n = 1,..,N \end{cases} \tag{4}$$

The equations of motion can be written in the following matrix form :

$$M\ddot{X} + C\dot{X} + K = M\ddot{X}_g \tag{5}$$

The normal frequencies are such as the solutions of the following system:

$$(-\omega^2 M + K)X = 0 \tag{6}$$

So

$$\det(-\omega^2 M + K) = 0 \tag{7}$$

We may express all normal frequencies relative to the same reference frequency which is ω_0, so that

$$\omega_n = \omega_0(1+\varepsilon)^{\frac{1}{2}} \quad with \quad n = 1,2,..,N \tag{8}$$

where:

$$\varepsilon = 2\beta\left(1 - \cos\left(\frac{n\pi}{N+1}\right)\right) \quad with \quad n = 1,2,...,N \tag{9}$$

The Taylor expansion of $\cos\left(\frac{n\pi}{N+1}\right)$ give the following approximate equation

$$\omega_i \approx \omega_0(1 + \frac{1}{2}\varepsilon) \tag{10}$$

In the next section, the optimization of the localized vibration energy in the mistuned oscillator system will be investigated by a multi-objective approach.

3 Multi-objective optimization problem

For the rest of the paper, we will consider a system involves eight oscillators coupled via linear springs while including two perturbations on the 2nd DOF and the 6th DOF.

$$\alpha_2 = \frac{m_2}{m} \text{ and } \alpha_6 = \frac{m_6}{m} \tag{11}$$

The problem of optimization of vibration energy localization in the quasi-periodic spring-mass system is formulated and solved as a combinatorial optimization problem.

The two mistuned DOFs response amplitudes are chosen as the objectives functions and the mistuning parameter, the coupling stiffness and the damping coefficient are the design variables.

The found approximations are used for solution of the problem by genetic optimization algorithms.

The optimization problem is posed as follows:

$$\begin{cases} \max\limits_{\alpha_2,\alpha_6,\beta,a} (X_{2\max}, X_{6\max}) \\ X_{2\max}(\alpha_2,\alpha_6,\beta,a) \\ X_{6\max}(\alpha_2,\alpha_6,\beta,a) \\ \alpha_{2\min} < \alpha_2 < \alpha_{2\max} \\ \alpha_{6\min} < \alpha_6 < \alpha_{6\max} \\ \beta_{\min} < \beta < \beta_{\max} \\ a_{\min} < a < a_{\max} \end{cases}$$

(12)

In order to study the localization phenomenon, the vibration localization criterion is defined as the difference between the amplitude of displacement of the mistuned DOF and that of the two neighboring DOFs.

$$\begin{cases} \tau_{(k-1)/k} = \dfrac{\left|X_{(k-1)\max} - X_{k\max}\right|}{\max(X_{(k-1)\max}, X_{k\max})} \\ and \\ \tau_{(k+1)/k} = \dfrac{\left|X_{(k+1)\max} - X_{k\max}\right|}{\max(X_{(k+1)\max}, X_{k\max})} \end{cases} \tag{13}$$

In the next section, the optimal parameters of the structure will be determined by the multi-objective optimization method and Pareto front.

4 Numerical results and discussion

For the rest of the paper, we will consider the design parameters listed in the following Table 1.

Table 1. Design parameters for the corresponding periodic structure

$m_n(kg)$	$k\ (N.m)$	$X_g(m)$
0.05	100	0.001

https://doi.org/10.1051/matecconf/201824101013

The intervals of variation of constraints of multi-objective problem are shown in Table 2.

Table 2. Intervals of variation of constraints of multi-objective problem

	α_2	α_6	β	a
B_{min}	1	1	0.005	0.01
B_{max}	1.1	1.1	0.015	0.02

4.1 Pareto front

The main objective of Pareto front is to find a compromise between the two amplitudes displacement in order to maximize the sum of energy harvesting.

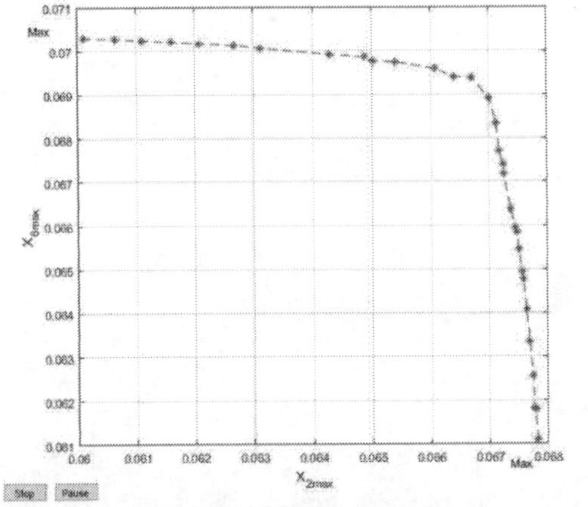

Fig. 2. Pareto front

We choose a point of the compromise zone:

$X_{2max} = 66.8 \times 10^{-3} m$ and $X_{6max} = 69.2 \times 10^{-3} m$

From this point, we look for the optimal values of the mistuning parameters α_2^* and α_6^*, the coupling stiffness β^* and the damping coefficient a^*.

We repeat the multi-objective study several times and from the presentation of the Pareto front we choose a point of the zone of compromise from which we determine the optimal parameters of the structure.

This approach yields the optimal intervals of the following structural parameters:

$$\begin{cases} 1.031 \le \alpha_2^* \le 1.041 \\ 1.039 \le \alpha_6^* \le 1.049 \\ 0.014 \le \beta^* \le 0.015 \\ 0.009 \le a^* \le 0.01 \end{cases} \quad (12)$$

4.2 Amplitude of displacement of mistuned DOFs

Figures 3-a and 3-b are obtained for the coupling stiffness (β^*=0.05) and the damping coefficient (a^*=0.01) and present the variation of the amplitude of displacement of mistuned DOFs as a function of the mistuning parameters (α_2 and α_6).

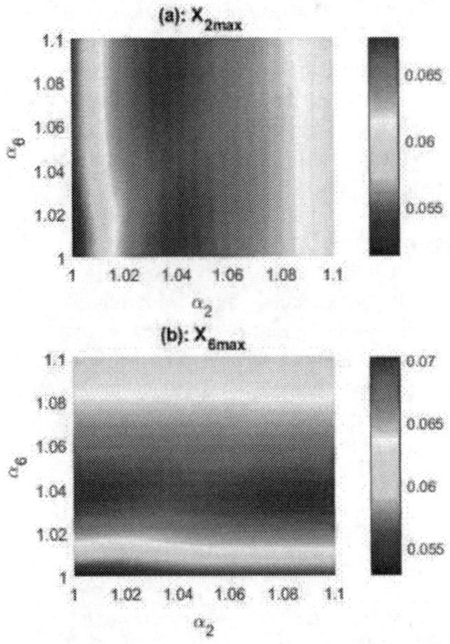

Fig. 3. Amplitude of displacement of mistuned DOFs according to mistuning parameters

Figure 3-a and 3-b show that around the optimal values of the mistuning parameters α_2 and α_6, the amplitude of displacement of the two mistuned DOFs is maximum.

4.3 Frequency response

For the two optimal mistuning parameters (α_2^*=1.036 and α_6^*=1.044), the coupling stiffness (β^*=0.05) and the damping coefficient (a^*=0.01), the frequency response of the system is shown in Figure 5

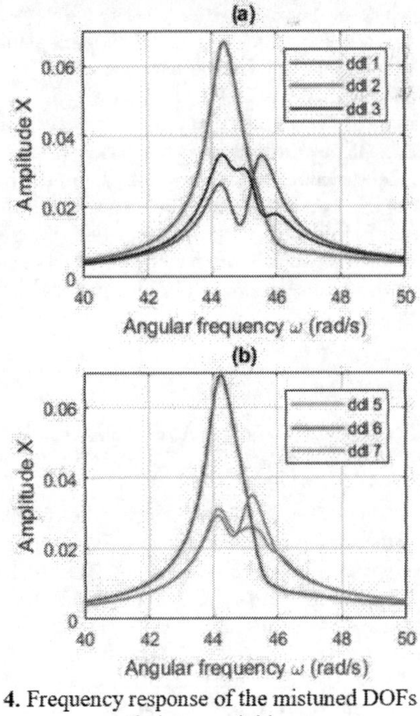

Fig. 4. Frequency response of the mistuned DOFs and of their two neighbors.

* Corresponding author: mariem.hbaieb@enis.tn

Figures 4-a and 4-b are obtained for the mistuning parameters (α_2^* =1.036 and α_6^* =1.044), the coupling stiffness (β^*=0.05) and the damping coefficient (a^*=0.01) showing the frequency responses of the mistuned DOFs and of their two neighbors.

These figures show that even two small mass perturbations at two DOFs of the periodic structure can lead to large difference in maximum amplitudes of oscillators.

It is also noted that the vibration energy will be located at the level of the mistuned mass with regard to a mass disturbance by adding matter (α is greater than 1).

This localized energy can therefore be harvested from harvesting devices placed at the level of the 2nd and the 6th oscillators.

The vibration localization criterion of the two DOFs mistuned and their neighboring DOFs is defined as:

$$\begin{cases} \tau_{2/1} = 48.22\% \quad and \quad \tau_{2/3} = 54.87\% \\ \tau_{6/5} = 47.89\% \quad and \quad \tau_{6/7} = 49.35\% \end{cases} \tag{13}$$

4 Conclusions

A generalized model of height horizontal oscillators coupled via linear springs while including two perturbations on two DOFs is developed. The objective of this study is to optimize the parameters of the quasi-periodic structure in order to maximize the localized vibration energy.

The problem of optimization of a quasi-periodic system mistuned is formulated and solved as a combinatorial optimization problem. An optimization method based on genetic optimization algorithm and Pareto front were developed for solving this problem.

In this paper, the parameters of a quasi-periodic system of coupled masses have been optimized by maximizing the two amplitudes of displacement of mistuned DOFs.

From the optimal parameters found, we verified the localization of the vibratory energy at the levels of the mistuned DOFs by the frequency response.

the periodic structures can be of very high dimension and which contains localized or distributed non-linearities.

The paper opens just what is happening in a small network. There is much more method of resolution when the network becomes narrower or in the presence of non-linearity.

References

1. C. Pierre and E. H. Dowell, Localization of vibrations by structural irregularity, *J. Sound Vib*, **114**, 19875 49–564
2. P. Malaji, S. Ali, Energy harvesting from near periodic structures, *J. Vibration Engineering and Technology of Machinery*, **23**, 411–420 (2015)
3. J. Hou and Ch. Cross, Minimizing Blade Dynamic Response in a Bladed Disk Through Design Optimization, *AIAA Journal*, **43**, 406-412 (2005)
4. E. Petrov, R. Vitali, and R. Haftka, Optimization of mistuned bladed discs using gradient-based response surface approximations, 41st Structures, Structural Dynamics, and Materials Conference and Exhibit, Structures, Structural Dynamics, and Materials and Co-located Conferences
5. Hong Hee Yoo, Jae Young Kim, Daniel J. Inman, Vibration localization of simplified mistuned cyclic structures undertaking external harmonic force, *Journal of Sound and Vibration*, **261**, 859–870 (2003)
6. P. Malaji, S. Ali, Analysis of energy harvesting from multiple pendulums with and without mechanical coupling, *The European Physical Journal Special Topics*, **224**, 2823–2838 (2015)
7. G. Litak, M. Friswell, C. Kwuimy, S. Adhikari, and M. Borowiec, Energy harvesting by two magnetopiezoelastic oscillators with mistuning, *theoretical & applied mechanics letters*, **2**, 043009 (2012)

A nonlinear model of the hand-arm system and parameters identification using vibration transmissibility

Nahid Hida, Mohamed Abid, and Faouzi Lakrad[a]

Laboratory of Renewable Energy and Dynamics of Systems, Faculty of Sciences Aïn Chock, University Hassan II-Casablanca, Casablanca, Morocco

Abstract. In the present paper a lumped single degree-of-freedom nonlinear model is used to study biodynamic responses of the hand arm system (HAS) under harmonic vibrations. Then, the harmonic balance method is implemented to derive the vibration transmissibility. Furthermore, Padé approximations are used in the identification process of biodynamic characteristics of the HAS model. This process is based on minimizing the distance between the theoretical and the experimentally measured transmissibilities.
The proposed identification workflow is applied to vibrations at the wrist in two cases: 1) the transmissibility versus the grip force for fixed excitation frequencies, and 2) the transmissibility versus the excitation frequency for fixed grip force.

1 Introduction

Lumped parameters linear models are widely used in the literature to study the biodynamic responses of the human hand-arm system (HAS) to vibration, see for instance [1] and [2]. These models range from single to multiple degrees of freedom mass-spring-damper systems. The corresponding coefficients are identified using fitting and optimization techniques that minimize the distance with measured mechanical impedances and/or vibration transmissibilities to different segments of the HAS.
In fact, these models impose constant stiffness, mass and damping coefficients while the skin-muscle-bone characteristics change with changing experimental conditions and their mechanical behaviors are knowledged to be nonlinear. Indeed, many factors highly influence the HAS vibration response, we cite among others, frequencies and amplitudes of the imposed vibrations, hand forces (the grip and the push-pull forces of the hand on the handle), the posture, the vibration direction, the handle geometry and the subject anthropometry, see [3] and [4].
Many contradictory results are reported in the literature because of the differences in the experimental conditions [3]. However, all experimental studies agreed that the transmissibility of the hand-arm segments show rapid decrease with increasing frequency and distance from the source of vibration [3]. Moreover, it is found that vibrations above 200 Hz are confined to the hand, while frequencies below 100 Hz and 40 Hz could be transmitted to the forearm and the upper arm, respectively.
In the present work, a nonlinear lumped single degree-of-freedom system is used to model the HAS. Then, the vibration transmissibility at the wrist is used to identify the parameters of this model under various conditions of the hand forces and the excitation frequencies. The main parameters of control of the HAS model are taken nonconstant and are expressed as fractional polynomials of the grip force or the excitation frequency.
The present paper is organized as follows: in section 2, a nonlinear mathematical model of the HAS is presented and an expression of the vibration transmissibility is derived. In section 3, the process of parameters identification is explained and applied to two case studies.

[a] e-mail: `f.lakrad@fsac.ac.ma`

2 Mathematical model

A mathematical single degree-of-freedom model of nonlinear vibrations of the HAS is shown in Fig.1. The viscoelastic characteristics of the HAS are modelled by restoring and dissipative forces that are assumed to be nonlinearly related to the displacement and the velocity, respectively. In what follows, cubic polynomial nonlinearities are considered.
The absolute displacement of the mass is denoted $x(t)$ and

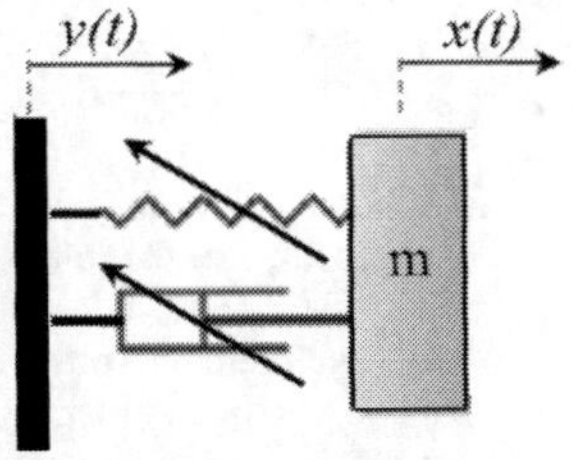

Fig. 1. The studied non-linear lumped hand-arm model. $y(t)$ is the imposed displacement at the handle.

the imposed harmonic displacement at the handle is denoted $y(t) = Y\cos(\Omega\, t)$, where t is time, Y and Ω are the amplitude and the frequency (measured in rad/s), respectively.
The nonlinear equation of motion of the mass is written as

follows

$$m\ddot{x} + c_1(\dot{x} - \dot{y}) + k_1(x - y) + c_3(\dot{x} - \dot{y})^3 + k_3(x - y)^3 = 0 \quad (1)$$

where k_1 and c_1 are the linear stiffness and viscous damping coefficients, respectively. The corresponding nonlinear coefficients are denoted k_3 and c_3.
A first order approximation of the periodic solution of Eq.(1), having the same frequency Ω as the imposed harmonic displacement of the handle, is obtained using the harmonic balance method (HBM) [5].
Thus, the solution is sought under the following form

$$x(t) = X\cos(\Omega t - \theta) \quad (2)$$

where X and θ are the amplitude and the phase of the absolute displacement. In Eq.(2) the contribution of higher order harmonics are assumed negligible.
The amplitude X is given by

$$X = \sqrt{Y^2 + a^2 + \frac{a^2(3k_3a^2 + 4k_1 - 4m\Omega^2)}{2m\Omega^2}} \quad (3)$$

where a is the the relative displacement amplitude of the mass. It is solution of the following sixth order algebraic equation

$$\frac{9}{16}\left(\Omega^6c_3^2 + k_3^2\right)a^6 + \frac{3}{2}\left(k_3(k_1 - m\Omega^2) + c_1c_3\Omega^4\right)a^4 + \left((k_1 - m\Omega^2)^2 + c_1^2\Omega^2\right)a^2 = m^2Y^2\Omega^4 \quad (4)$$

It is worth noting that nonlinearities i.e., k_3 and c_3 are causing the existence of polynomial terms of order 4 and 6 in Eq. (4) and are leading to the possibility of coexistence of three periodic solutions.
The transmissibility of the displacement is given by

$$TR = \frac{X}{Y} = \sqrt{1 + \frac{a^2}{Y^2}\Big(\frac{3k_3a^2 + 4k_1 - 2m\Omega^2}{2m\Omega^2}\Big)} \quad (5)$$

Combining Eqs. (4) and (5), the transmissibility can be expressed as

$$TR = \frac{a}{mY\Omega^2}\sqrt{\Big(\frac{3}{4}k_3a^2 + k_1\Big)^2 + \Omega^2\Big(\frac{3}{4}\Omega^2c_3a^2 + c_1\Big)^2} \quad (6)$$

In order to use the derived nonlinear transmissibility given in Eqs. (5) or (6), the amplitude a of the relative displacement should be measured. Otherwise, it becomes an unknown of the optimization problem and physically meaningful constraints should be imposed to identify it.
The linear case is obtained by cancelling the effects of nonlinearities i.e., setting $k_3 = c_3 = 0$. Hence, the amplitude of the linear relative displacement is

$$a_L = \frac{mY\Omega^2}{\sqrt{(k_1 - m\Omega^2)^2 + c_1^2\Omega^2}} \quad (7)$$

Consequently, the classical linear transmissibility TR_L is given by

$$TR_L = \sqrt{\frac{k_1^2 + c_1^2\Omega^2}{(k_1 - m\Omega^2)^2 + c_1^2\Omega^2}} \quad (8)$$

It is worth pointing out that the linear transmissibility (8) is independent of the imposed handle motion amplitude Y, while the nonlinear transmissibility (6) is dependent on it. Based on vibration measurements on the HAS, Adewusi et al. [3] found experimentally, using two different magnitudes of broad-band random vibrations of the handle, that the excitation magnitude affects the transmissibility, particularly around the characteristic frequencies.

3 Parameters identification

The model parameters identification process is based on experimental measurements of the HAS vibration transmissibility, along the forearm direction i.e., Z_h axis, that were performed on nine subjects under various hand forces and excitation frequencies [4].
Experiments were conducted for a hand-arm posture corresponding to bent-arm with 90° elbow angle. Pan et al. [4] used a six-degree polynomial to fit the experimental transmissibility in terms of the grip force. Then, they computed the mean transmissibility of the nine subjects.
In the present paper, two case studies of the wrist vibration transmissibility are investigated. For each case the transmissibility is measured while varying a parameter of control. Then, in each case a parameter of the model is chosen to be expressed in terms of the parameter of control.
In the first case study, the control parameter that is varied is the grip force F_g. Specifically, four discrete sinusoidal vibrations with frequencies 10Hz, 16Hz, 25Hz and 40Hz are used as imposed displacement on a cylindrical handle with a constant velocity of 0.1 m/s. The corresponding imposed accelerations on the handle-arm interface are, respectively, 6.28 m/s^2, 10.05 m/s^2, 15.71 m/s^2 and 25.13 m/s^2.
In this first case study, the linear transmissibility model (8) is used and the linear stiffness k_1 is expressed in terms of the grip force F_g using the following Padé approximation

$$k_1 = \frac{\alpha_0 + \alpha_1F_g + \alpha_2F_g^2 + \alpha_3F_g^3}{\beta_0 + \beta_1F_g + \beta_2F_g^2} \quad (9)$$

where $\alpha_i, i = 0..3$ and $\beta_j, j = 0..2$ are constants to be identified.
In the second case study, the transmissibility is measured while varying the control parameter that is the excitation frequency, for fixed values of hand forces. The nonlinear transmissibility model (6) is used in the process of identification. The relative displacement amplitude a is expressed in terms of the excitation frequency Ω using the following fractional polynomials form

$$a = \frac{a_0 + a_1\Omega + a_2\Omega^2 + a_3\Omega^3}{a_4 + a_5\Omega + a_6\Omega^2} \quad (10)$$

In both cases the model parameters are identified by minimizing the distance between the theoretical formulas of the transmissibility (8) or (6) and the measured mean transmissibility at the wrist [4].
Furthermore, since the results of the minimization problem are not unique and not surely global, additional physically meaningful constraints are to be imposed to the parameters of the model. Hence, the following constraints are imposed on the linear parameters

$$0 < m \leq 3\ [\text{Kg}]\ ;\ 50 < c_1 < 500\ [\text{N.s/m}]\ ;\ k_1 > 0\ [\text{N/m}]$$

The imposed constraint on the relative displacement amplitude a, used in the second case study, is

$$0 \leq a \leq Y\,(1 + TR\,) \tag{11}$$

In the present work, the nonlinear stiffness and damping coefficients are not explicitly expressed in terms of the control parameters.

3.1 Application of the linear model

The nine parameters of the linear model to be identified are the mass m, the linear damping coefficient c_1 and the seven parameters, α_i and β_i, defining the linear stiffness k_1 given in Eq. (9).
The outcome of the optimization process, in the case of 10 Hz excitation, is that the identified mass and damping coefficient are m = 2.45 Kg and c_1 = 104.34 N.s/m. The expression of the linear stiffness in terms of the grip force is given by

$$k_1 = \frac{0.175053 - 0.024855F_g - 0.005343F_g^2 + 9.0\times10^{-6}F_g^3}{0.000016 - 0.000008F_g - 1.49\times10^{-7}F_g^2}$$

The corresponding sum of squares error (SSE), between the theoretical and the measured transmissibilities, is equal to 0.0016. This low value of SSE proves the high accuracy of the theoretical model.
Figure 2 shows a good agreement between the computed linear transmissibility (represented in lines) and the experimental transmissibilities (in symbols) versus the grip force for four excitation frequencies (in different colors). Figure 2 shows that increasing the excitation frequency leads to decreasing the transmissibility at the wrist. Moreover, for each frequency the transmissibility increases with increasing the grip force till reaching a saturation-like phenomenon or starting to decrease. The increasing transmissibility phase is longer with increasing the excitation frequency.
In Fig.3 are shown the predicted natural frequency ω and

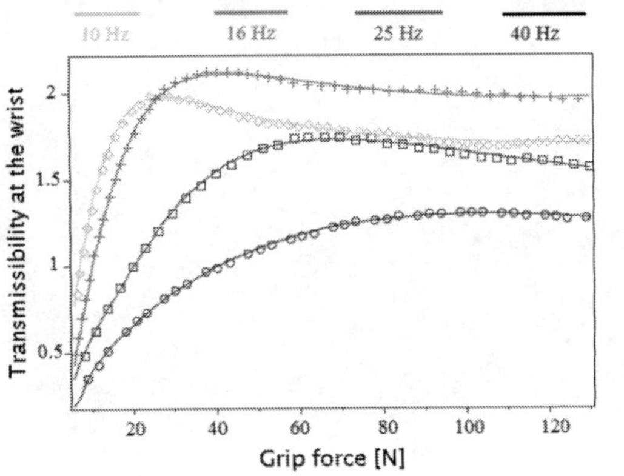

Fig. 2. The transmissibility at the wrist versus the grip force for various excitation frequencies. The lines correspond to the theoretically computed transmissibility. The measured data [4] are shown in symbols (◇ + □, o).

the damping ratio ξ versus the grip force for 10 Hz sinusoidal frequency excitation. They are defined as follows

$$\omega = \sqrt{\frac{k_1}{m}} \quad \text{and} \quad \xi = \frac{c_1}{2m\omega} \tag{12}$$

Figure 3 shows that the natural frequency is increasing and the damping ratio is decreasing with increasing the grip force. In fact, increasing the grip force rigidifies the system and consequently increases the natural frequency. Moreover, the increase rate of the natural frequency is decreasing with increasing F_g.

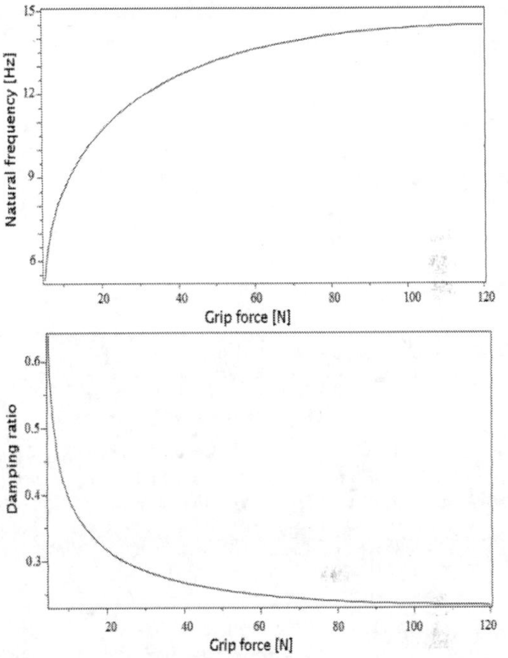

Fig. 3. Variation of the natural frequency and the damping ratio versus the grip force for the excitation frequency 10 Hz.

3.2 Application of the nonlinear model

In this subsection, the developed nonlinear transmissibility (6) is studied versus the excitation frequency for a fixed grip force and for varying amplitude Y. Indeed, the linear transmissibility (8) is independent of the handle displacement amplitude Y. Consequently, changing Y will not affect the curve of the transmissibility versus any parameter of control.
Experimentally, the transmissibility was measured, at various locations of the HAS, versus the excitation frequency for various values of the hand forces in [4]. The imposed velocity amplitude V at the handle was taken constant equal to 0.012 m/s, and the corresponding imposed displacement amplitude is $Y = V/\Omega$.
Since the relative displacement amplitude a was not measured in [4], then six parameters of the nonlinear model are to be identified : the amplitude a, the mass, the linear and nonlinear stiffness and damping coefficients.

Furthermore, in order to perform the optimization process, the amplitude a is expressed as a fractional polynomial of the excitation frequency Ω, see Eq.(10).
Figure 4 shows the good fit between the results of the nonlinear transmissibility given in Eq. (6) and the experimentally measured transmissibility at the wrist for 75N grip force [4].
Figure 4 shows a resonant frequency of the wrist, between 12Hz and 17Hz, that corresponds to the maximum of vibration transmissibility. Moreover, the transmissibility is amplified, i.e., higher than one, for excitation frequencies below 60 Hz. Consequently, low frequencies vibrations are a potential danger for the wrist.

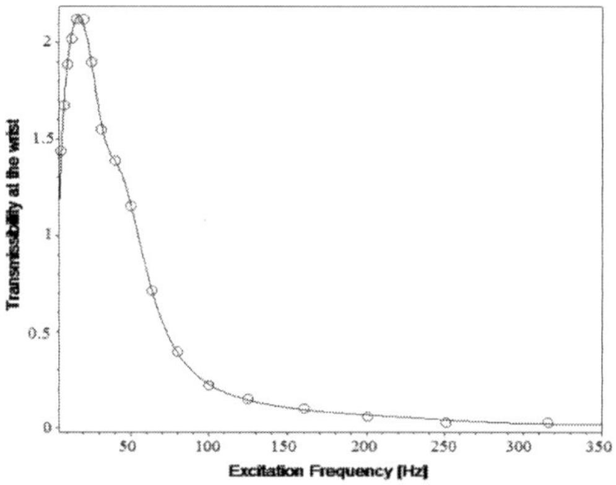

Fig. 4. Transmissibility at the wrist versus the excitation frequency for 75 N grip force. Experimental data [4] are shown in circles and the theoretical results in continuous line.

4 Conclusion

In the present study a nonlinear lumped single degree-of-freedom system is used to model the hand arm system response under sinusoidal imposed displacement of the handle. Then, the vibration transmissibility is derived analytically using the harmonic balance method.
Furthermore, the process of the model parameters identification is based on the minimization of the distance between measured and theoretical transmissibilities.
The derived workflow is applied to two case studies: 1) the transmissibility is measured versus the grip force; and 2) the transmissibility is measured versus the excitation frequency.
In order to perform the identification process the linear stiffness is expressed in terms of the grip force in the first case. In the second case, the amplitude of the relative displacement of the mass is taken dependent on the frequency. In both cases, the Padé approximations are used.
The outcome of the present work is that varying parameters models are fitting better to experimental measurements than the constant parameters models. Furthermore, the nonlinear model is introduced and it fits the case where the magnitude of the imposed displacement is varied.
As a continuation of the present work the mass, the stiffness and the damping coefficients will be allowed to vary in terms of influencing factors on the vibration transmissibility. Moreover, the proposed identification process will be extended to higher d.o.f. models.

References

1. Rakheja S, Wu JZ, Dong RG, Schopper AW. *A comparison of biodynamic models of the human hand-arm system for applications to handheld power tools*. Journal of Sound and Vibration **249** : 55-82, 2002.
2. Dong RG, Welcome DE, Xu X, Chen Q, Lin H, McDowell TW, Wu JZ. *A model for simulating vibration responses of grinding machineworkpiece- hand-arm systems*. Journal of Sound and Vibration **431** : 276-294, 2018.
3. Adewusi SA, Rakheja S, Marcotte P, Boutin J. *Vibration transmissibility characteristics of the human hand-arm system under different postures, hand forces and excitation levels*. Journal of Sound and Vibration **329** : 2953-2971, 2010.
4. Pan D, Xu XS, Welcome DE, McDowell TW, Warren C, Wu J, Dong RG. *The relationships between hand coupling force and vibration biodynamic responses of the handarm system*. *Ergonomics*, DOI: 10.1080/00140139.2017.1398843, 2017.
5. Nayfeh AH. *Perturbation methods*. Wiley-VCH Verlag; 2004.

Application of vectorcardiography and recurrence-based methods to analysis of ECG signals

Joanna Iwaniec[1,*], *Marek* Iwaniec[2] and *Antoni* Kalukiewicz[3]

[1]AGH University of Science and Technology, Faculty of Mechanical Engineering and Robotics, Department of Rob otics and Mechatronics, Mickiewicz Alley 30, 30-059 Kraków, Poland
[2]AGH University of Science and Technology, Faculty of Mechanical Engineering and Robotics, Department of Process Control, Mickiewicz Alley 30, 30-059 Kraków, Poland
[3]AGH University of Science and Technology, Faculty of Mechanical Engineering and Robotics, Department of Mining, Dressing and Transport Machines, Mickiewicz Alley 30, 30-059 Kraków, Poland

Abstract. The paper concerns application of vectorcardiography and recurrence-based methods to analyze ECG signals. The first method involves recording the direction and magnitude of electrical forces of the heart by means of a continuous series of vectors that form a curving line around a center, while the second consists in representing the ECG signals in the phase space and searching for the recurrence properties of the registered signals. In both methods, change of analysis domain offers new opportunities for early diagnosis. In this work, discussed methods have been used for the purposes of analysis of selected ECG signals stored in the medical database. Vectorcardiograms were determined on the basis of ECG signals for the Farnk's leads. Signal analysis was carried out with the application of the recurrence plots (RP) method. The proposed approach simplifies assessment of the signal information content by providing patterns characteristic for different myocardial injury. Carried out research requires complementation by results of statistical analysis for various age groups and degrees of myocardial injury.

1 Introduction

For many years, by far the greatest threat to the health and life all over the World have been diseases of the circulatory system. The basic and the most efficient tool used in cardiovascular diagnostic is ECG, which is so widespread that it is difficult to determine the exact number of these tests performed each day. Despite the requirement for the efficient and reliable methods of ECG signals analysis and interpretation, in practice, almost exclusively, classical methods involving analysis of geometrical properties of the recorded waveforms in time domain are used. The scientific works carried out recently have been focused on increasing the accuracy of ECG analysis by increasing the accuracy of finding its characteristic points. However, the accuracy limits of such a geometrical analysis have already been reached. Therefore the quest of formulating novel methods supporting classical analysis of ECG signals is still open.

In this work, in order to support classical methods of ECG analysis, vectorcardiography and recurrence-based methods have been used. The first method involves recording the direction and magnitude of electrical forces of the heart by means of a continuous series of vectors that form a curving line around a centre. The main idea of the second method consists in representing the ECG signals in the phase space and searching for the recurrence properties of the registered signals.

In both methods, change of analysis domain offers new, so far unreachable, opportunities for early diagnosis. In this work, discussed methods have been used for the purposes of analysis of selected ECG signals stored in the medical database *The PTB Diagnostic ECG Database*. Patients were selected from the group of male smokers after the age of 50. Vectorcardiograms were determined on the basis of ECG signals for Farnk's leads. The further analysis was carried out with the application of the recurrence plots (RP) and RQA analysis methods.

The structure of the paper is as follows. Section 2 concerns electrocardiography and vectorcardiography. Recurrence-based methods are discussed in the section 3. In the section 4 selected experimental results are presented. Finally, the paper is concluded in section 5.

2 Vectocardiography

Currently, electrocardiography is one of the most extensively described procedures of modern medicine. Ease of implementation, low costs and a relatively high diagnostic value make it a tool useful for every clinician. The electrodes on the patient's body are positioned in such a way that they project the heart work on two planes - the front one (three limb leads - aVR, aVL, aVF) and transverse (six precordial leads, from V_1 to V_6). In addition, it is possible to extend the set of electrodes to revise cardiac areas that are not available in

* Corresponding author: jiwaniec@agh.edu.pl

the routine testing. Typical result of the test has the form of a printout of voltage in the function of time for all the leads mentioned above, additionally extended by bipolar leads, which are called differential leads since they register differences between selected unipolar leads. Lead I presents the difference between aVL and aVR, lead II - between aVF and aVR while III – between aVF and aVL. They do not add any new information to the electrocardiogram data space, but only improve its interpretation by the clinician. A lot of basic diagnostic information can be derived based on the signal from lead II only.

Vectorcardiography [1] presents slightly different approach. Since the working heart behaves like a dipole, it is possible to analyse changes of a dipole moment in time. The idea of the method consists in conversion of one-dimensional information from each lead into three coordinates in the Cartesian coordinate system, which reduces the dimensions of the data space from originally nine to three. Visualization of a vectorcardiogram is usually performed using specialistic software. It has the form of a curve determined by a time-wandering vector of the dipole moment of the heart. The method is very promising since it shows the changes of direction and magnitude of the electrical forces of the heart [2, 3]. This information is of key importance for the diagnosis of acute coronary syndromes. Unfortunately, in clinical practice, classical methods are still preferred. The hope for practical applications of vectorcardiography lies in the usage of numerical methods.

3 Recurrence-based methods

The formal concept of recurrences, in the sense of returning to the former states, was introduced by Henri Poincaré [4] in 1890. In 1987, based on the Poincaré's achievements, Eckmann, Oliffson and Ruelle formulated the **R**ecurrence **P**lots (RP) method [5]. The method aims at revealing all the times when the phase space trajectory of the considered dynamical system visits roughly the same area in the phase space. Recurrence of given state x (observed at time i) at a different time j is marked with ones and zeros within so called recurrence matrix $[R]$:

$$[R_{i,j}] = \Theta\left(\varepsilon - \left\| \{x_i\} - \{x_j\} \right\|\right) \quad i, j = 1, \ldots, N \qquad (1)$$

where N denotes the number of considered states x_i, ε_i threshold distance, $\| \ \|$ a norm and $\Theta(\ )$ the Heaviside function.

Graphical representation of matrix $[R]$ is known under the term *recurrence plot*. In other words, the recurrence plot represents graphically the matrix $[R_{i,j}]$ given as:

$$[R_{i,j}] = \begin{cases} 1: & \{x_i\} \approx \{x_j\} \\ 0: & \{x_i\} \neq \{x_j\} \end{cases} \quad i, j = 1, \ldots, N \qquad (2)$$

where $\{x_i\} \approx \{x_j\}$ are the points belonging to the neighbourhood of radius ε (defined according to the applied norm).

In order to estimate recurrence plot for the given dynamical system it is necessary to specify values of three parameters: threshold ε, time delay τ and embedding dimension m. Values of these parameters have a significant influence on the informative content of the estimated recurrence plot and, therefore, should be selected carefully.

In the literature various criterions for selection of threshold ε have been proposed [6, 7, 8]. The most commonly used criterion is the fixed amount of nearest neighbours (FAN) criterion, which consists in fixing the number of neighbours for each point of the trajectory [9]. In this case, the threshold is different for each point of the trajectory. The advantage of the method is that it preserves the recurrence point density and allows to compare RPs of different systems without the necessity of normalising the time series beforehand.

Appropriate time delay τ can be determined with the application of the auto-correlation or mutual information function [10, 11]. For the purposes of the smallest sufficient embedding dimension m estimation, the false nearest neighbours method [10] is frequently used. Originally, the RP method was used for the purposes of visualisation of system trajectories. Later, development of the **R**ecurrence **Q**uantification **A**nalysis (RQA) [12], supporting interpretation of recurrence plots, has consolidated the RP method as a tool for nonlinear data analysis.

4 Results of the experimental research

Results of the research carried out over the last few years [14, 15, 16] have proven that the recurrence-based methods are particularly sensitive to changes in the dynamic behaviour of mechanical and biological systems. In this paper, the methodology based on RPs method was applied to analysis of selected ECG signals stored in the medical database *The PTB Diagnostic ECG Database*. The database contains 549 records from 290 subjects (patients). Each subject is represented by one to five records. Each record includes 15 simultaneously measured signals: the conventional 12 leads together with the 3 Frank lead ECGs (vx, vy, vz). Each signal is digitized at 1000 samples per second, with 16 bit resolution over a range of ± 16.384 mV.

In this paper, the proposed methodology is presented at the example of ECG signal of 66-year-old patient (patient no. 9 in the Database), performed 11 days after the infarction of the lower-lateral wall.

Patient vectorcardiograms are presented in the Fig. 1. In the ECG recording (Fig. 2) the most important changes are ST segment elevation in leads V_2, V_3, V_4, negative T wave visible in lead V_6 and pathological Q wave in leads II and III, which is characteristic for the infarction of the lower-lateral wall.

In the next step of analysis, ECG signals from Frank leads (vx, vy, vz) were analysed by means of the recurrence plot method (Fig. 3). Minimal dimensions of the phase space m were determined with the application of the false nearest neighbours criterion. In order to establish time delay τ mutual information function method was used. For the purposes of consistency, for all the analysed signals the same value of threshold ($\varepsilon = 0.1$) was assumed. All the computations were carried out in the CRP Toolbox for MATLAB.

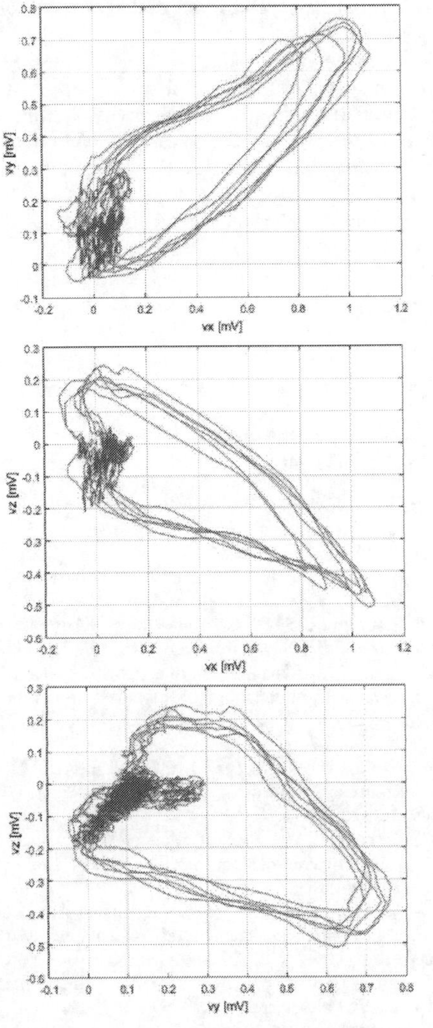

Fig. 1. Vectorcardiogram of the considered patient.

In the next step of analysis, conversion of analysed ECG signals from cartesian to spherical coordinates was performed:

$$r = \sqrt{x^2 + y^2 + z^2}, \quad \phi = arctg\frac{y}{x}, \quad \theta = \arcsin\frac{z}{r} \qquad (3)$$

Recurrence plots illustrating recurrent properties of the analysed signals in the spherical coordinates are presented in the Fig. 4.

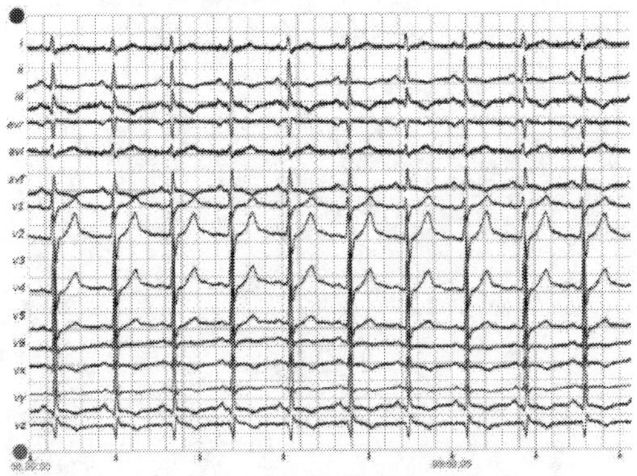

Fig. 2. ECG recording of the considered patient.

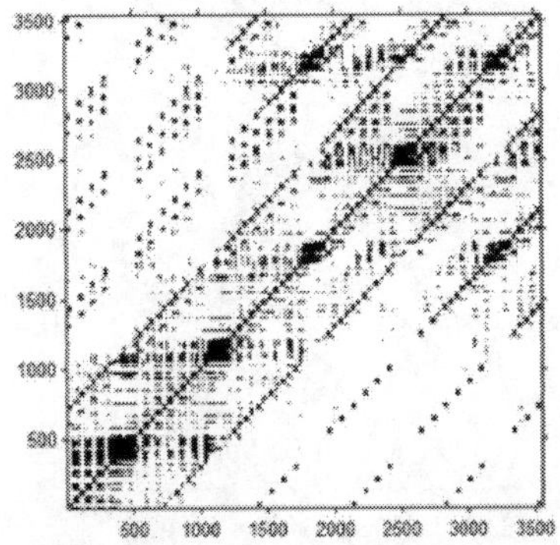

a) **vx**: $\tau = 90$, $m = 6$, $\varepsilon = 0.1$, FAN

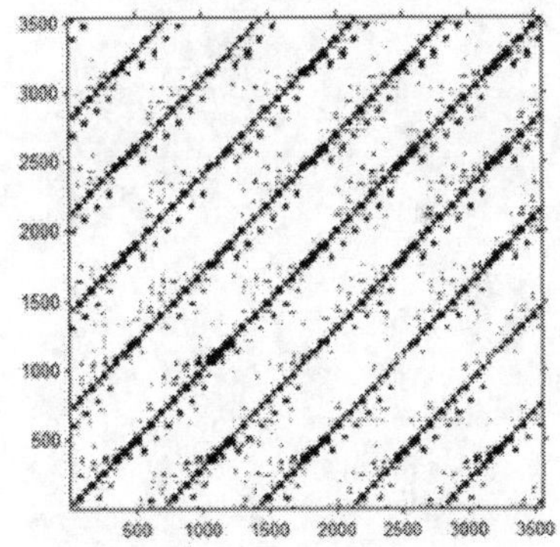

b) **vy**: $\tau = 90$, $m = 6$, $\varepsilon = 0.1$, FAN

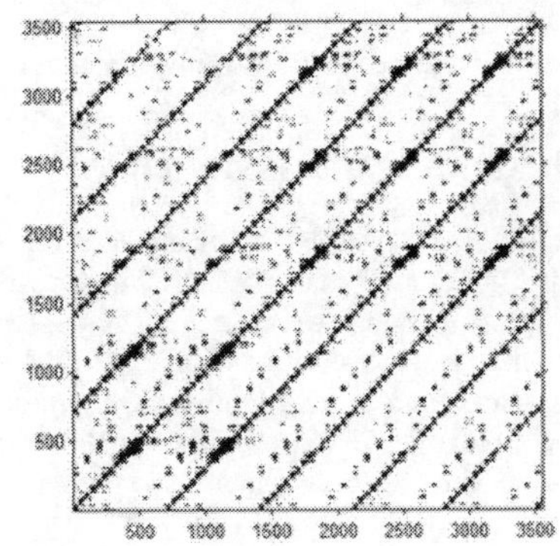

c) **vz**: $\tau = 90$, $m = 6$, $\varepsilon = 0.1$, FAN

Fig. 3. RPs of ECG signals from Frank leads.

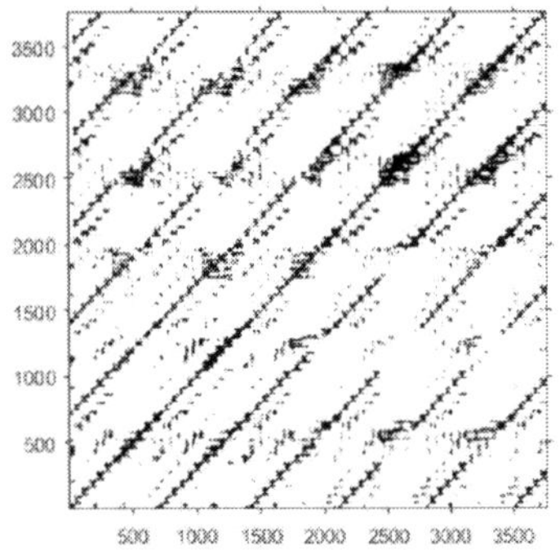

a) crp(*r*): $m = 5$, $\tau = 60$, $\varepsilon = 0.1$, FAN

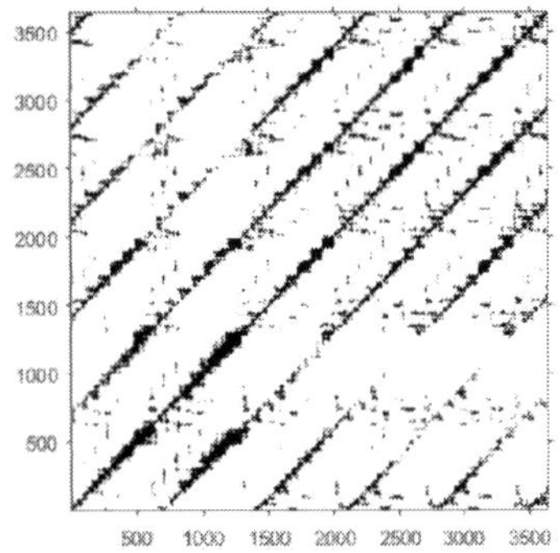

b) crp(θ): $m = 5$, $\tau = 90$, $\varepsilon = 0.1$, FAN

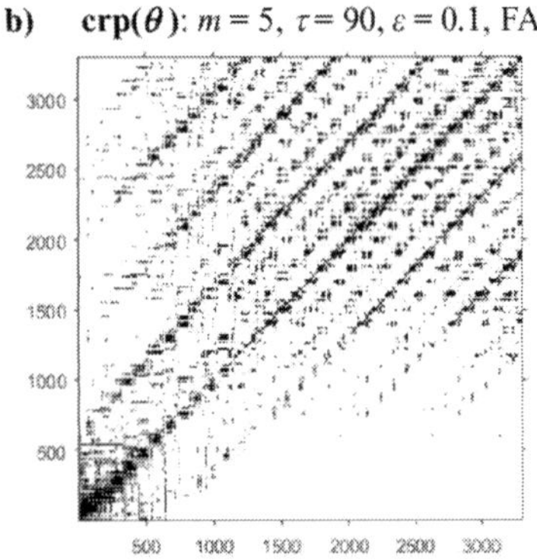

c) crp(φ): $m = 8$, $\tau = 100$, $\varepsilon = 0.1$, FAN

Fig. 4. RPs of ECG signals from Frank leads represented in the spherical coordinates.

The proposed approach simplifies assessment of the signal information content by providing patterns characteristic for different cardiovascular dysfunctions.

5 Conclusions and final remarks

In order to support classical methods of ECG signal interpretation in time domain, vectorcardiography and recurrence plots method were used. The essence of vectorcardiography consists in reduction of dimensions of the data space from originally nine (one-dimensional information from nine Einthoven leads) to three (three coordinates in the Cartesian coordinate system). The advantage of the method is the ability to observe the changes in the cardiac rhythm, which is of key importance in the diagnosis of acute coronary syndromes. However, such representation is difficult to interpret for cardiologists. Therefore, for the purposes of simplifying interpretation of information carried by the ECG signals from Frank leads (*vx*, *vy*, *vz*) application of the recurrence plots method was proposed. Selection of RP method is natural due to the cyclic character of the heart work

Conducted studies have to be continued and complemented by statistical analysis for different age groups and various degrees of severity of dysfunctions. However, at the present stage of the research, it can be concluded that application of the recurrence based methods can be successfully used for ECG rhythm analysis and for interpreting the results of vectorcardiography.

References

1. E. Frank, An Accurate, Clinically Practical System For Spatial Vectorcardiography, Circulation **13** (1956), 737–749.
2. G.E. Burch, The history of vectorcardiography, Med. His. Sup. **5** (1985), 103–131.
3. G. Daniel, G. Lissa, D. Medina Redondo, L. Vásquez, D. Zapata, Real-time 3D vectorcardiography: An application for didactic use, J. Phys. Conf. Ser. 90 (2007), doi:10.1088/1742-6596/90/1/012013.
4. H. Poincaré, Sur la probleme des trios corps et les equations de la dynamique, Acta Math. **13** (1890), 1-271.
5. J.P. Eckmann, K.S. Oliffson, D. Ruelle, Recurrence plots of dynamical systems, Europhys. Lett. **5** (1987), 973-977.
6. M. Thiel, M.C. Romano, J. Kurths, R. Meucci, E. Allaria, F.T. Arecchi, Influence of observational noise on the recurrence quantification analysis, Physica D **171**, 3 (2002), 138-152.
7. N. Thomasson, T.J. Hoeppner, C.L. Webber, J.P. Zbilut, Recurrence quantification in epileptic EEGs, Phys. Lett. A **279**, 1-2 (2001), 94-101.
8. J.P. Zbilut, C.L. Weber, Embedding and delays as derived from quantification of Recurrence Plot, Phys. Lett. A **171** (1992), 199-203.
9. N. Marwan, M.C. Romano, M. Thiel, J. Kurths, Recurrence plots for the analysis of complex systems, Phys. Rep. **438** (2007), 237-329.
10. H. Kantz, T. Schreiber, *Nonlinear Time Series Analysis* (University Press, Cambridge, 1997).
11. N. Marwan, *Cross Recurrence Plot Toolbox for Matlab. Reference Manual. Version 5.15. Release 28.6* (available at: http://tocsy.pik-potsdam.de, 2010, accessed: 10 January 2015).
12. J.P. Zbilut, C.L. Weber, Embedding and delays as derived from quantification of Recurrence Plot, Phys. Lett. A **171** (1992), 199-203.
13. E.J. Ngamga, A. Nandi, R. Ramaswamy, M.C. Romano, M. Thiel, J. Kurths, Recurrences of strange attractors. PRAMANA-J. Phys. **70**, 6 (2008), 1039-1045.
14. J. Iwaniec, P. Kurowski, Experimental verification of selected methods sensitivity to damage size and location, J. Vib. Control **23**, 7 (2017), 1133-1151.
15. J. Iwaniec, Investigation of selected mechanical systems by recurrence plots method, Int. J. Struct. Stab. Dy. **13**, 7 (2013).
16. J. Iwaniec, M. Iwaniec, Heart work analysis by means of recurrence-based methods, Diagnostyka **18**, 4 (2017), 89-96.

Nonlinear Dynamics of a Planar Hinged-Supported Beam with One End Lumped Mass and Longitudinal Elastic Support

Lukasz Kloda[1,2*], *Stefano* Lenci[1], and *Jerzy* Warminski[2]

[1]Department of Civil and Buildings Engineering, and Architecture, Polytechnic University of Marche, Ancona, Italy
[2]Department of Applied Mechanics, Lublin University of Technology, Lublin, Poland

Abstract. Nonlinear forced-damped oscillations of a non-slender hinged simply supported beam with mass and spring attached to one end are investigated by mean of a finite element method. The frequency response curves are constructed numerically and the variability of hardening/softening behaviour of frequency response curves due to the lumped mass and axial linear spring stiffness is investigated. Resonant and sub resonant motion of beam midpoint as well as jumps between solution branches are highlighted.

1 Introduction

In mechanical systems undergoing incidental or operational excitation, a special attention must be paid to nonlinear dynamic phenomena which must be well understood by mechanical and structural engineers. Nonlinear dynamics effects may arise especially in the vicinity of natural frequencies. Currently at the initial designing stage a real structure can be adequately modelled numerically e.g. by finite element method and then examined to understand and optimize the way it works, without building advanced laboratory test devices or sophisticated analytical algorithms. Moreover, such numerical analysis is available to cover predictable global phenomena in a wide range as well as unexpected local effects, see [1,2]. The main disadvantages of this approach is studying only specific cases and needing huge computational time for very complex systems. However, a development of computing machines increases significantly possibilities of numerical models enabling to detect special effects which are difficult to measure. It causes that unreachable things for today can be common in the near future.

Recently a few papers have been published on coupled transversal and longitudinal oscillations of beams with different boundary conditions. Application of passive control by arbitrary assumed end mass and varying one end stiffness of a spring kinematically excited in the longitudinal direction is presented in [3]. In [4] authors focused on parametrically excited beam-axial spring-lumped mass system. However, the ratio of the beam mass and tip mass was assumed as a very small number thus the study was limited to a particular case.

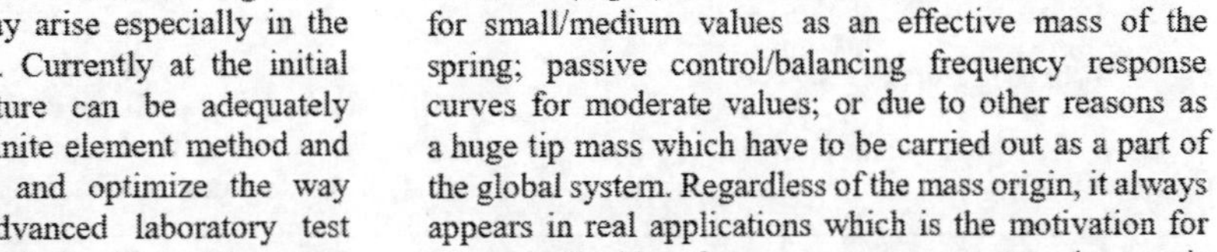

Fig. 1. The beam-spring-tip mass system. Thick curve - deformed beam, thick dashed line - rest configuration.

In this paper we are focused on numerical computations of a planar beam-spring system, wherein the effect of the additional end mass is investigated. Namely, a hinged-simply supported beam with a tip mass and a linear spring axially restraining one end of the beam is studied (Fig. 1). The added end mass can be understood for small/medium values as an effective mass of the spring; passive control/balancing frequency response curves for moderate values; or due to other reasons as a huge tip mass which have to be carried out as a part of the global system. Regardless of the mass origin, it always appears in real applications which is the motivation for research. Besides, frequency response curves hysteresis and an associated jump phenomenon will be addressed in this work.

2 Finite element model

The finite element model of the structure consists of 100 straight equal length beam elements of *B31*-type with linear geometric order; linear spring of stiffens k_s; tip mass M_t subjected to one end which undergoes only translatory inertia in the longitudinal direction (mass rotation is neglected). Beam characteristic parameters are collected in Table 1. The corresponding boundary conditions are:

* Corresponding author: l.kloda@pollub.pl

$$U(0,T) = U(L,T) = 0, \quad (1)$$

$$M(0,T) = M(L,T) = 0, \quad (2)$$

$$W(0,T) = 0, \quad (3)$$

where Eq. (2) corresponds to vanishing of bending moments at beam ends. The right end side of the beam is ruled by the balance equation in the axial direction (see Fig. 2):

$$N(L,T)\cos\varphi + V(L,T)\sin\varphi + k_s W(L,T) - M_t \ddot{W}(L,T) = 0. \quad (4)$$

where: N - normal force, V - shear force, k_s - spring stiffness, M_t - tip mas, φ - slope angle, L - beam length in initial unrestrained rest configuration. A dot means derivative with respect to time T.

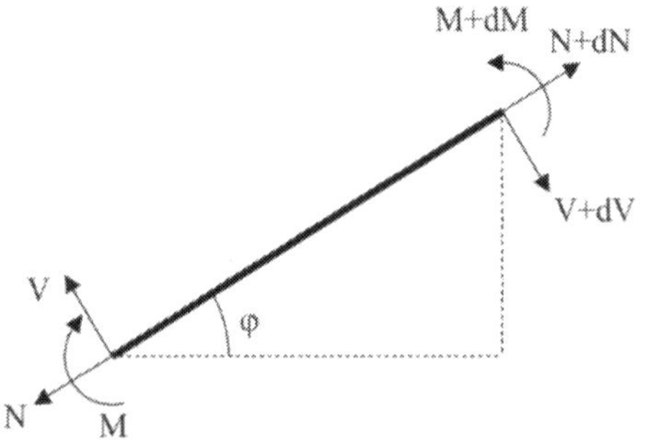

Fig. 2. Deformed beam element – coplanar forces.

Table 1. Beam dimensions and system mechanical properties [5].

Initial length [L]		500 mm
Cross section [A]		50 mm × 50 mm
Density [ρ]		7850 kg/m^3
Young modulus [E]		210 Gpa
Poisson's ratio [ν]		0.3
Beams shear factor [Sk]		0.85
Viscous damping factor [ζ]		6%
Tip mass [M_t]		$0 \leq M_t < \infty$ (0, 1 kg, 2 kg 1000 kg)
Spring stiffness $\kappa = \frac{k_s L}{EA}$	simply supported end	$\kappa = 0$
	free to move with end spring	$0 < \kappa < \infty$ ($\kappa = 1$)
	hinged	$\kappa = \infty$
Excitation amplitude [p_v]		40.7992 kN

It is worth to remark that for zero spring stiffness the system with tip mass is free to move, for k_s equals to infinity the joint becomes independent of the end mass, restrained (typically hinged). Between those two limit values of stiffness ($0 < k_s < \infty$) the system turns out to be free to move-spring-tip mass boundary value problem in time domain.

End mas and spring stiffness do not influence the transversal linear natural bending modes, but they affect the longitudinal natural frequencies and corresponding linear modes. However, for larger deformations those modes are coupled and then due to nonlinear dynamics of the beam motion is much more complex.

To study forced oscillations, in simulations, the beam midpoint has been excited in vertical direction (in X- direction, see Fig. 1) by a periodic force of amplitude p_v which corresponds to 1 mm static deflection of hinged simply supported beam and circular frequency $\alpha\omega_0$:

$$P(T) = p_v \cos(\alpha\omega_0 T), \quad (5)$$

where: ω_0 is natural frequency of the first bending mode; α is multiplicative detuning coefficient.

In order to obtain the frequency response curves, α has been varied in the range from 0.8 to 1.125, around the first natural frequency ω_0 (461.47 Hz, 2899.50 rad/s), step by step forward as well as backward, by disregarding the transient behaviour and keeping only the steady state solution. More precisely, the amplitude of steady state transversal oscillations of the node $Z = L/2$ has been recorded with sampling at each $1/40$ of the period of frequency excitation $\alpha\omega_0$. The motion in the longitudinal direction, which in general has double frequency of lateral oscillations, has been recorded, too.

3 Results

At the beginning the time histories of the recorded point have been analysed only for selected excitation frequencies, but even those preliminary simulations have brought some phenomena which need deeper investigations. One of them is the effect of the tip mass on the trajectory of the beam midpoint, which undergoes (as expected) coupled motion because axial inertia (of both beam and tip mass), as confirmed by the shifts the intersection of symmetry axis ($U = 0$). In Fig. 3a we show that for small tip mass $M_t = 2$ kg trajectories are moved toward right of about 0.00132 mm for $\alpha = 0.8$ and 0.00812 mm for $\alpha = 1.1$. Fig. 3b, on the other hand, presents huge oscillations (21.45 mm) for extremely big tip mass (1000 kg), which cause orbits shifts toward left of about 0.265 mm and $0.00775 - 0.0021$ mm for $\alpha = 0.8$ and $\alpha = 1.1$, respectively. Furthermore, for huge mass the orbit is bend right, differently to what happens with small or no tip mass. This proves that the end mass can significantly change dynamics of the beam, and it is necessary to carry out a deeper analysis.

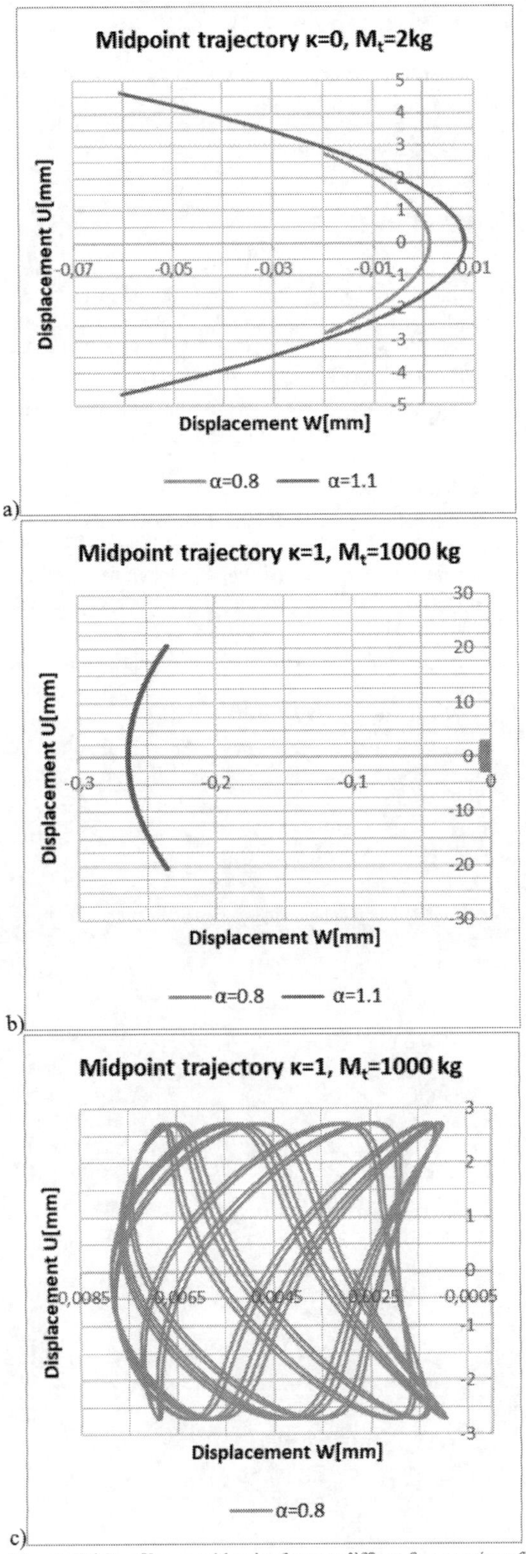

Fig. 3. Orbits of beam mid point for two diffrent frequencies of excitation $\omega_n = 0.8$ and $\omega_n = 1.1$ for spring stiffnesses: (a) $\kappa = 0$, $M_t = 1kg$, (b, c) $\kappa = 1$, $M_t = 1000$ kg.

For system parameters $\kappa = 1$, $M_t = 1000$ kg and $\alpha = 0.8$, a multifrequency response has been detected (see Fig. 3c, which is a zoom of Fig. 3b). Motion of the recorded node is composed of two frequencies $\omega = 230.47$ Hz (1448.09 rad/s) and $\omega = 738.28$ Hz (4638.75 rad/s), which correspond to the frequency of the first longitudinal mode $\omega = 230.26$ Hz (1446.77 rad/s) and frequency of the excitation $\omega = 0.8\ \omega_0 = 738.35$ Hz (4639.19 rad/s), respectively. This is another relevant dynamical property of the considered system, which could be better analysed in the future.

The numerically obtained Frequency Response Curves (FRC) are given in Fig. 4. As already presented in [2] for the beam with no additional mass, the FRC in the absence of the end spring is softening, while as the end spring stiffness increases the system becomes hardening up to the limit point of restrained end ($\kappa = \infty$). Additional small end mass causes mollification of characteristics, e.g. for simply supported beam with $M_t = 2$ kg (20,38 % of beam mass) the behaviour is significantly more softening, while for $\kappa = 1$, $M_t = 1$ kg (10,19 %) it has the same tendency with negligible differences. The interesting thing is that a beam with huge mass $M_t = 1000$ kg has hardening nature. Besides, the additional mass affects the simply supported beam more than restrained by the spring. The examined cases give a motivation to deeper analytical works in the further investigations.

The well-known jump phenomenon has been observed, both for the softening behavior ($\kappa = 0$, $M_t =$ 2 kg and $\kappa = 0$, $M_t = 0$ kg) as well as for the hardening one (in this case the upper jump points cannot be seen in Fig. 4 due to the limited range of excitation frequency, but they can be easily foreseen).

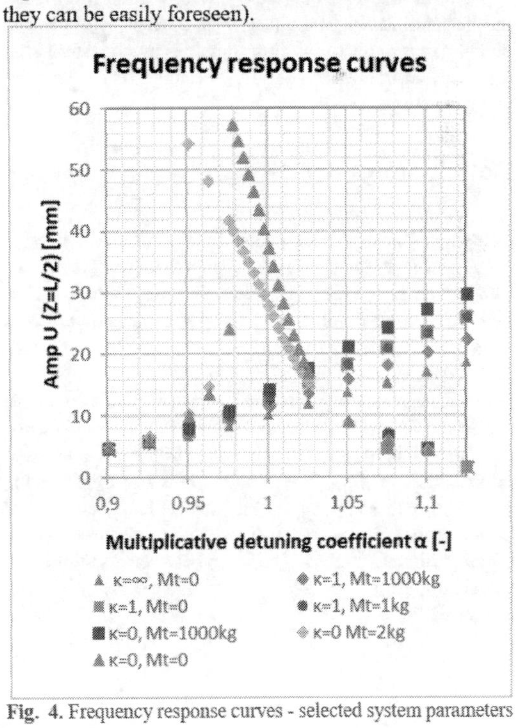

Fig. 4. Frequency response curves - selected system parameters

Midpoint trajectory κ=1, M_t=1000 kg

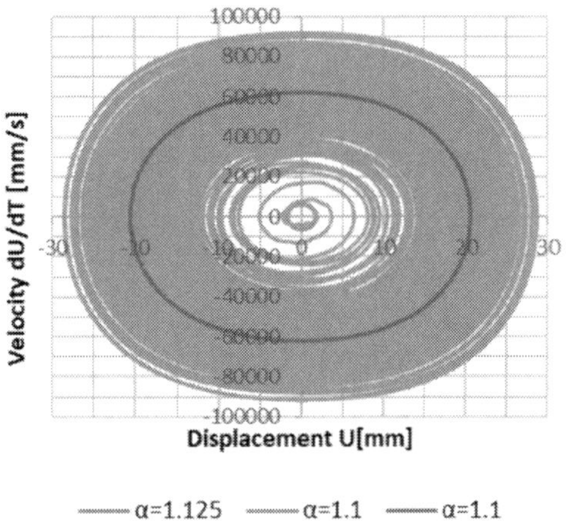

Fig. 5. Trajectories across the jump between large and small amplitude solutions.

This jump phenomenon is now better illustrated. Figure 5 presents, in the phase space $(U, \dot{U})$, the trajectories of the midpoint across the jump phenomenon. The internal small ellipse (orange color) is a picture of the steady state small amplitude oscillation for $\alpha = 1.125$. Through a gradual decrement of the excitation frequency (from $\alpha = 1.125$ to $\alpha = 1.1$) the solution has been smoothly perturbed, then the trajectory went out from original steady state up to the other large amplitude oscillation, represented by external ellipse (blue color). The widest area (grey color) is the long transient needed for the stabilization of the system on the new solution. For higher damping factor motion stabilizes much faster in the aftermath of the jump, although it changes the frequency response curve.

Conclusions

The paper shows some aspects of the nonlinear dynamics of selected cases of a simply supported beam-spring-lumped mass system, modelled by the finite element method (the commercial software Abacus_CAE® is used). The influence of the tip mass on nonlinear forced-damped oscillations of the beam has been examined. Two scenarios have been considered, with (a) relatively small ratio tip mass to beam mass 10 - 20 % (1-2 kg) and for (b) a huge added mass 10 000 % (1000 kg). It enables to control the frequency response curves by mean of the added tip mass, which can be useful in practical applications. However, changes are more essential for small spring stiffnesses. The jump between solutions and coexisting trajectory stabilization have been observed numerically.

Acknowledgement

This work is part of the collaboration between Polytechnic University of Marche and Lublin University of Technology, which is aimed at developing a Joined Doctoral Programme.

References

1. L. Kloda, S. Lenci, J. Warminski, Nonlinear dynamics of a planar beam–spring system: analytical and numerical approaches, J. Nonlinear Dyn. PDF avaliable online at https://doi.org/10.1007/s11071-018-4452-2 (2018) 1-18.
2. L. Kloda, S. Lenci, J. Warminski, Nonlinear dynamics of a planar hinged-supported beam with one end spring system, MATEC Web of Conferences **148** (2018) 06004.
3. A. Shibata, S. Ohishi, H. Yabuno, Passive method for controlling the nonlinear characteristics in a parametrically excited hinged-hinged beam by the addition of a linear spring, J. of Sound Vib. **350** (2015) 111-122.
4. A. Aurami, H. Yabuno, Cubic–quintic nonlinear parametric resonance of a simply supported beam, J. Nonlinear Dyn. **90** (2017) 549-560.
5. Abaqus 6.14 Documentation available at http://dsk.ippt.pan.pl/docs/abaqus/v6.13/index.html (2013).

Using the Gottwald and Melbourne's 0-1 test and the Hugichi fractal dimension to detect chaos in defective and healthy ball bearings

C. A. Kitio Kwuimy[1,*], *T.* Haj Mohamad[2,**], and *C.* Nataraj[2,***]

[1]Department of Engineering Education, College of Engineering and Applied Science, University of Cincinnati, Cincinnati, OH 45221

[2]Villanova Center for Analytics of Dynamic Systems (VCADS), Villanova University, Villanova, PA 19085, USA

Abstract. The paper considers the identification of chaotic behavior dynamics using the data extracted from an experimental model of rotor supported on rolling elements. A description of the method is provided as well as an illustration using a standard dynamic map. The 0-1 test for chaos and the Higuchi dimension are shown to be effective tool in the identification of chaotic behavior of the system bearing with and without faults.

1 Introduction

Time series analysis is generally referred to as the transformation of observed dynamic data into insight for making better decisions [1]. The data extracted from the behavior of dynamic systems can be non-periodic and eventually stochastic. For specific applications these data need to be characterized and analyzed in order to extract information about the system. Several methods have been used in various branches of science including engineering, epidemiology, biology and finance [2–4]. Traditional methods for time series in these domains include statistical analysis and dynamic method which include methods in the time domain, methods in the state space domain and methods in the frequency domain.

The most used frequency analysis method is the Fast Fourier Transform (FFT), which is in fact a linear method based on the analysis of the spectrum of the frequencies in the response of the system. For instance, a periodic system will be completely described by its FFT while for a non-periodic or stochastic system, the broad-banded FFT will give no specific information [4]. State space methods include all methods developed in the state space of the system. These methods involve all state variables of the system dynamics. The key point here is the necessity of the state space, which in general (when one cannot obtain the measurement of the dynamics variables of the system) needs to be reconstructed from one of the available measurement. State space methods include the Lyapunov exponent, traditional fractal dimensions such as the correlation dimension, and the method of recurrence analysis. Although these methods have been widely used [2–4], the pitfall here is that they require excessive computing in terms of reconstruction of the state space and some of these methods (Lyapunov exponent for example) are very sensitive to noise and required a certain level of advanced expertise from the user [3, 4, 7]. The time domain methods which consist of a simple plot of the data as function of time. Such plot provides preliminaries information about the nature (periodic, stochastic, chaotic) of the data. Additionally, statistical measurement and fractal measurement of the time response have been shown to provide reliable and efficient information about the nonlinear nature of the system.

The later is considered in this paper. We consider a machine fault simulator and collect data for three conditions of the system (healthy and defective with different levels of severity). The objective is to provide additional insight in the system response. In fact, in previous contribution, we used statistical methods and methods of the state space, namely the method of recurrence analysis for similar investigations [8, 9, 11–

*e-mail: cedrick.kwuimy@uc.edu
**e-mail: thajmoha@villanova.edu
***e-mail: c.nataraj@villanova.edu

14]. Extension was made to the detection and classification of faults. This paper is focussed on identification of chaotic response. The 0-1 test for chaos and the Higuchi's fractal dimension are used to analyze the time series. The methods have been previously used in various branches of science, including finance, engineering and biology [7, 10].

The paper is organized as follows. In Section 2, the description of the methods is summarized. In Section 3, data from healthy and defective experimental systems are analyzed, periodic and chaotic responses are considered. The last section is the conclusion.

2 Methods of analysis

2.1 The 0-1 test for chaos

The 0-1 test for chaos was developed by Gottwald and Melbourne [7] to distinguish between regular and chaotic dynamics in deterministic dynamical systems. Rather than requiring phase space reconstruction which is necessary to apply standard chaotic indicators, the test works directly with the time series and does not involve any preprocessing of the data. The method is based on the estimation of a statistical parameter K which is used to conclude on the periodic ($K \to 0$) and chaotic ($K \to 1$) nature of the a time signal. To compute K from a time series

$$X = x(1), x(2), ...x(N) \tag{1}$$

where N is the total number of data collected, the time series is projected into a plane (p, q) defined as

$$p(n,c) = \sum_{j=1}^{n} \phi(j) \cos jc, \quad q(n,c) = \sum_{j=1}^{n} \phi(j) \sin jc \tag{2}$$

where $n = 1, 2, .., N$ and $c \in (0, 2\pi)$ is an arbitrary constant. The diffusive and non-diffusive behavior of $p(n, c)$ and $q(n, c)$ are used to estimate K. For this end, the adjusted mean-square displacement is estimated. It is given by [7]

$$\begin{aligned} D(n,c) &= \frac{1}{N}\sum_{j=1}^{N}\left[\Delta_1^2 + \Delta_2^2\right] - \Phi^2 \frac{1-\cos nc}{1-\cos c} \\ \Delta_1 &= p(n+j,c) - p(j,c), \\ \Delta_2 &= q(n+j,c) - q(j,c) \\ \Phi &= \frac{\lim}{N \to \infty} \frac{1}{N}\sum_{j=1}^{N} \phi(j) \end{aligned} \tag{3}$$

To ensure the limit $N \to \infty$, in general the limit $n \leq N_0$ with $N_0 \ll N$ ($N_0 = N/20$ in this paper) is considered. The second term in Eq. (3) is added to minimize eventual fluctuation in $M(n, c)$. For each values of c, the parameter K_c is estimated as

$$K_c = \frac{\text{cov}\,(\zeta, \delta)}{\sqrt{\text{var}(\zeta)\text{var}(\delta)}} \tag{4}$$

where $\zeta = 1, 2, .., N_0$ and $\delta = D(c, 1), D(c, 2), ..D(c, N_0)$. The final value of K is estimated as the median of all $K_c = f(c)$. The use of the median instead of the average value is justified by the fact that, it avoid large drift caused by possible resonance in a specific value of c. See Refs. [7] for illustrations.

2.2 Higuchi fractal dimension

The Higuchi's fractal dimension is a powerful fractal measurement of the structure of data [10]. The specificity here is that, unless other fractal dimensions which are estimated in the (reconstructed) state space, the Huguichi's fractal dimension can be computed from a single time series. The Higuchi's fractal dimension has been used as dynamic tool in several applications, including mechanical engineering and geophysics [10].

For the time series in Eq. (1), one can extract different short time series as

$$X_k^m = x(m), x(m+k), ...x\left(m + \left\lceil \frac{n-m}{k} \right\rceil k\right) \tag{5}$$

where $\lceil . \rceil$ is the ceiling function. The length of the curve associated to each new time series X_k^m is then computed as the sum of algebraic distances between consecutive elements of the new time series, that is

$$L_m(k) = \sum_{i=1}^{\lceil \frac{n-m}{k} \rceil} \left(x(m+ik) - x(m-(i-1)k)\right) \tag{6}$$

The dimension is computed using the following scaling relation

$$< L(k) > \propto k^{-D} \tag{7}$$

where $< L(k) >$ is the average lengths of the segment used to measure the total length of the time series at scale k, and $L(k)$ is the normalized length defined as

$$L_m = \frac{N-1}{\lceil \frac{n-m}{k} \rceil k} \sum_{i=1}^{\lceil \frac{n-m}{k} \rceil} \left(x(m+ik) - x(m-(i-1)k)\right) \tag{8}$$

The Higuchi fractal dimension is then estimated as the slope of the least square linear fit of $\langle \text{Ln}\,(L(k)) \rangle = \text{Ln}\,(1/k)$

3 Detection of chaotic behavior in bearings

3.1 The experimental setup and data collection

The machine fault simulator used as experimental setup is shown in Fig. 1. It consists of a motor with a closed loop speed control connected to a shaft supported on rolling element bearings. The system can be loaded using a mass and balancing disks. A load mass of 5 kg is used in this research. The balancing disk has two rows of screw holes drilled near the edge into which different masses can be inserted to balance the shaft. Data are captured with three different bearings. Two of these bearings have an outer race defect with different fault size and one bearing does not have any defect. The signals from the experimental setup are collected with accelerometers placed at the bearing case.

Figure 1. Experimental setup

3.2 Data analysis and evidence of chaos

In this section, data are collected at speed 40 Hz for the healthy and defective conditions. Bearings with outer race defects have impulses excited by the rolling element entering the defect. Because of these impulses there is a significant difference between the signals from bearings with and without defects.

The $(p(n,c), q(n,c))$ projection of the data are shown in Fig. 2 for the system with small defect. The diffusive structure of the plot is clearly a signature of a non-deterministic dynamics. For different values of c, the global behavior of Fig. 3 is obtained and for this specific set of parameters, the corresponding median is quite identical ($K = 0.99$) for the three states of the system as shown in the figures.

The graphs of Fig. 3 are obtained using the procedure for described above for the Higuchi dimension. The slopes are obtained respectively as $H = 1.96$ for the healthy system, $H = 1.81$ and $H = 1.95$ for the

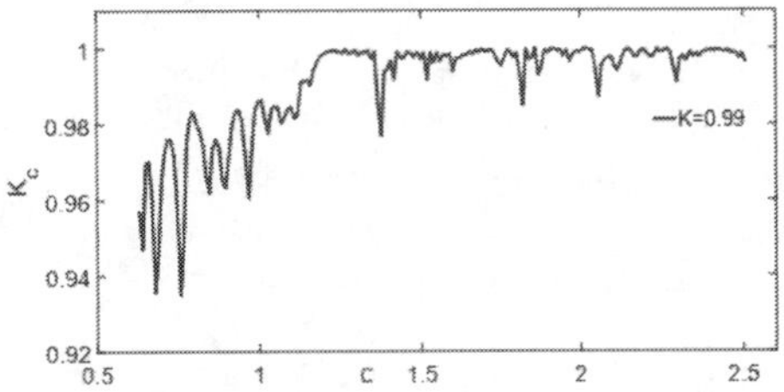

Figure 2. Estimation of the 0-1 test parameter of the bearing system with small defect

system with small and large defects. The figure also shows an extension of the procedure for different set of data corresponding to rotor speed 50 Hz and 10 Hz. The results are structurally characteristic of a non-periodic response.

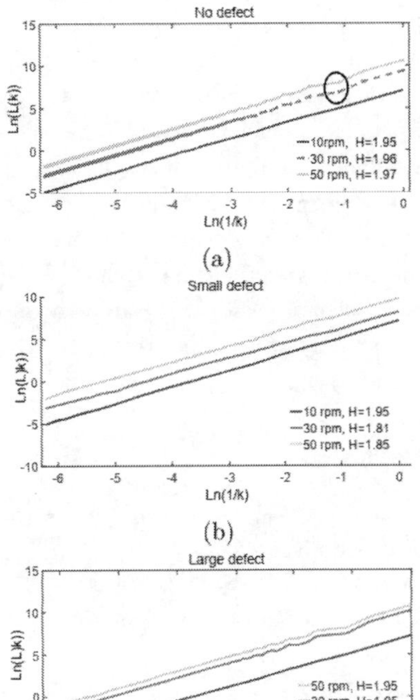

Figure 3. Estimation of the Higuchi fractal dimension. (a): Healthy system. (b): System with small defect. (c): System with large defect.

As in previous contributions [8], the structure of the time response of the bearing can be quite complex. This complexity was initially captured through the structure of the recurrence plots matrix and quantify using the so-called recurrence quantification parameters. The results obtained in this paper, provide a more "simplistic" identification of chaos. The Higuchi dimension can be seen here as a graphical tool (from which a quantification can be deduced) and it is complement with the 0-1 test.

4 Conclusion

The main objective of this work was to identify chaotic response of bearing system with and without defects. Two method were used for this main. The 0-1 test for chaos the Higuchi fractal dimension. The data were analyzed and the methods have been shown to be an effective tool for the identification of chaotic behavior.

The results obtained in this paper are extensions of results obtained in other applications, where the 0-1 test and the Higuchi dimension were used either separably to analyze experimental data. One key relevance of this work is the relative simplicity of the methods and their robustness. Thus, they appear as strong candidate for the characterization of experimental data as well as fault analysis. In fact, in these two applications, simpler and robust methods are always advantageous (computational cost and robustness). Current investigations are carried out to analyze the effects of noise and the dynamics of the system under various parameter configurations.

Acknowledgments

Part of this work was done during the Postdoctoral fellow of C.A.K. Kwuimy at the VCADS. He would like to thank Dr. C. Nataraj and the US Office of Naval Research for the financial support (grant N00014-13-1-0485)

References

[1] E.A. Boyd, Analytics **7**, 6 (2012)

[2] A. Galka, *Topics in Nonlinear Time Series Analysis* (World Scienti, 2000)

[3] H. Kantz, T. Schreiber, *Nonlinear Time Series Analysis* (University Press-Cambridge, 2004)

[4] H.D.I. Abarbanel, *Analysis of Observed Chaotic Data* (Springer-Verlag, New York, 1996)

[5] C.A.K. Kwuimy, C. Nataraj, *Structural Nonlinear Dynamics & Control and Diagnosis* (Springer-Verlag, 2015), pp. 97–123

[6] N. Marwan, M.C. Romano, M. Thiel, J. Kurths, Physics Reports **438(5-6)**, 237 (2007)

[7] A.G. Gottwald, I. Melbourne, SIAM Journal of Applied Dynamical Systems **8**, 129 (2009)

[8] C.A.K. Kwuimy, M. Samadani, K. Kappaganthu, C. Nataraj, in *International Conference on Vibration Engineering and Technology of Machinery* (2014)

[9] T. Haj Mohamad, M. Samadani, C. Nataraj, Journal of Vibration and Acoustics, **140(6)**, 061009 (2018).

[10] T. Higuchi, Physica, **23**, 277 (1988)

[11] M. Samadani, T. Haj Mohamad, C. Nataraj, *Feature Extraction for Bearing Diagnostics Based on the Characterization of Orbit Plots With Orthogonal Functions*, in *ASME 2016 International Design Engineering Technical Conferences and Computers and Information in Engineering Conference*, pp. V008T10A026–V008T10A026

[12] T. Haj Mohamad, C.A.K. Kwuimy, C. Nataraj, *Discrimination of Multiple Faults in Bearings Using Density-Based Orthogonal Functions of the Time Response*, in *ASME 2017 International Design Engineering Technical Conferences and Computers and Information in Engineering Conference*, pp. V008T12A035–V008T12A035

[13] T.H. Mohamad, Y. Chen, Z. Chaudhry, C. Nataraj, Journal of Software Engineering and Applications **11**, 181 (2018)

[14] T. Haj Mohamad, A.A. Cavalini, V. Steffen, C. Nataraj, *Detection of Cracks in a Rotating Shaft Using Density Characterization of Orbit Plots*, in *Proceedings of the 10th International Conference on Rotor Dynamics – IFToMM*, edited by K.L. Cavalca, H.I. Weber (Springer International Publishing, Cham, 2019), pp. 90–104, ISBN 978-3-319-99268-6

Analysis of non symmetric FG sandwich plates under Thermo-mechanical loading using a novel shear deformation theory with stretching effect

H. Saidi, A. Bouchafa, A. Tounsi, A. E.A. Adda.Bedia

Laboratoire des Matériaux et Hydrologie, Université de Sidi Bel Abbes, Département de Génie Civil et des Travaux Publics, Faculté de Technologie, Algérie.

Abstract. The analysis of non-symmetric functionally graded sandwich plates under thermo-mechanical loading is developed using a novel hyperbolic shear deformation theory and considering thickness stretching effects. This theory accounts for adequate distribution of the transverse shear strains in the thickness of the plate and satisfies the traction free boundary conditions on the top and bottom surface of the plates, thus a shear correction factor is not required. The governing equations of equilibrium of non-symmetric functionally graded sandwich plates can be obtained using virtual work principle and the closed form solutions are obtained by using Navier technique. The accuracy of the present results is established by comparing those with well known trigonometric shear deformation theories. The results are presented for deflections and stresses of non-symmetric simply supported square plates.

1 Introduction

Composite laminated materials are finding wide application in many engineering fields. Therefore, a number of approximate analytical and numerical methods have been developed in this area. Such composites are made of two or more materials to obtain a good structural performance, which the constituent does not show individually. Recently, advanced composite materials known as functionally graded material have attracted much attention in many engineering applications due to their advantages of being able to resist high temperature gradient while maintaining structural integrity [1]. The functionally graded materials (FGMs) are microscopically inhomogeneous, in which the mechanical properties vary smoothly and continuously from one surface to the other. They are usually made from a mixture of ceramics and metals to attain the significant requirement of material properties. FGMs are widely used in many structural applications such as mechanics, civil engineering, optical, electronic, chemical, mechanical, biomedical, energy sources, nuclear, automotive fields and ship building industries to eliminate stress concentration and relax residual stresses and enhance bond strength.

Several studies have been performed to analyze the behavior of functionally graded plates and shells. Reddy [2] analyzed the static behavior of functionally graded rectangular plates based on his third-order shear deformation theory of plates.

Cheng and Batra [3] related the deflections of a simply supported functionally graded polygonal plate given by the first-order shear de formation theory and a third-order shear de formation theory to those of an equivalent homogeneous Kirchhoff plate. Cheng and Batra [4] also presented results for the buckling and steady-state vibrations of a simply supported functionally graded polygonal plate based on Reddy's plate theory. Loy et al. [5] studied the vibration of functionally graded cylindrical shells using the Love shell theory. Employing the sinusoidal shear formation theory (SSDT),

Zenkour [6] presented the Navier analytical solution for FG plates.

Li et al. [7] developed a three dimensional solution for free vibration of multi-layer FGM system-symmetric and unsymmetric FGM sandwich plates using the Ritz method. However, various higher-order shear deformation theories are developed using five unknown functions. The well-known higher-order plate theories with five unknown functions are as follows: (i) parabolic shear deformation plate theory (PSDPT) [8], (ii) sinusoidal shear deformation plate theory (SSDPT) [9], and (iii) exponential shear deformation plate theory (ESDPT) [10].

In this study, a novel higher order shear deformation theory for linear analysis of moderately thick functionally graded sandwich plate is examined. In this model the displacement varies as a hyperbolic function across the plate thickness and satisfies zero shear stress condition at the top and bottom surfaces of the plate, thus a shear correction factor is not required. The accuracy of the present results is established by comparing those with well known trigonometric shear deformation theory.

2 Problem formulation

Let us consider the case of a sandwich plate composed of three microscopically heterogeneous layers as shown in Fig. 1. Rectangular Cartesian coordinates are used to describe infinitesimal deformations of a three-layer sandwich elastic plate occupying the region [0, a] · [0, b] · [_h/2, +h/2] in the unstressed reference configuration. The mid-plane of the composite sandwich plate is defined by z = 0 and its external bounding planes being defined by z=±h/2, the equations will be derived in tensorial notations and specialized afterwards for the problem under consideration. The top and bottom layers of non-symmetric sandwich plate are made of an isotropic homogeneous material ceramic and metal, respectively, while the core layer is made of a mixture of (ceramic and metal) materials.

The effective material properties for each layer, like Young's modulus and Poisson' s ratio, can be expressed as:

$$P^{(n)}(z) = P_m + (P_c - P_m)V^{(n)} \quad (1)$$

where P_m and P_c denote the property of the bottom and top faces of layer 1, respectively, and vice versa for layer 3 depending on the volume fraction $V^{(n)}$(n = 1,2,3). Note that P_m and P_c are, respectively, the corresponding properties of the metal and ceramic of the FGM sandwich plate.

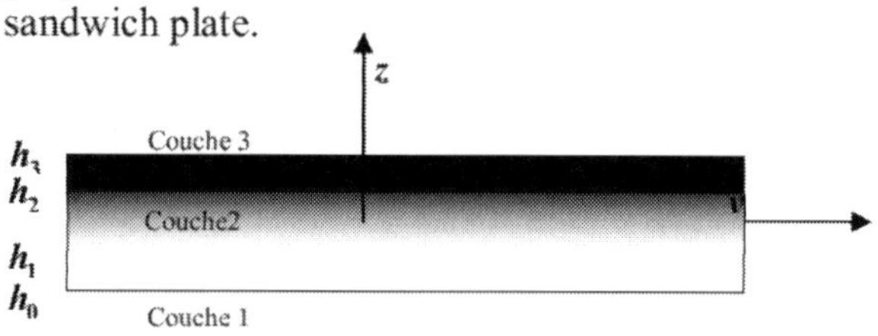

Figure1. Geometry of the FGM sandwich plate.

3 Governing equations for FG sandwich plate

The displacement field for FG sandwich plates is described in the following equations:

$$u(x,y,z) = u_0(x,y) - zw_{,x} + f(z)\theta_x \quad (2a)$$
$$v(x,y,z) = v_0(x,y) - zw_{,y} + f(z)\theta_y \quad (2b)$$
$$w(x,y,z) = w_0(x,y) + \acute{f}(z)\theta_z \quad (2c)$$

where, u, v, w are displacements in the x, y, z directions, u_0, v_0 and w_0 are midplane displacements, θ_x, θ_y and θ_z rotations of the *yz*, *xz*, and *xy* planes due to bending, respectively. $f(z)$ represents the shape function determining the distribution of the transverse shear strains and stresses along the thickness. It is given by

$$f(z) = \frac{\cosh(\pi/2)}{[\cosh(\pi/2)-1]} z - \frac{(h/\pi)\sinh\left(\frac{\pi}{h}z\right)}{[\cosh(\pi/2)-1]} \quad (2d)$$

with $\theta_z \neq 0$.

The linear strain expressions derived from the displacement model are as follows:

$$\begin{Bmatrix} \varepsilon_x \\ \varepsilon_y \\ \gamma_{xy} \end{Bmatrix} = \begin{Bmatrix} \varepsilon_x^0 \\ \varepsilon_y^0 \\ \gamma_{xy}^0 \end{Bmatrix} + z \begin{Bmatrix} k_x \\ k_y \\ k_{xy} \end{Bmatrix} + f(z) \begin{Bmatrix} \eta_x \\ \eta_y \\ \eta_{xy} \end{Bmatrix} \quad (3a)$$

$$\begin{Bmatrix} \gamma_{yz} \\ \gamma_{xz} \end{Bmatrix} = f'(z) \begin{Bmatrix} \gamma_{yz}^{0} \\ \gamma_{xz}^{0} \end{Bmatrix} \tag{3b}$$

Using the principle of virtual work, the following expressions can be obtained:

$$\int_{-h/2}^{h/2} \int_{\Omega} \left[\sigma_x \delta \varepsilon_x + \sigma_y \delta \varepsilon_y + \sigma_z \delta \varepsilon_z + \tau_{xy} \delta \gamma_{xy} + \tau_{yz} \delta \gamma_{yz} + \tau_{xz} \delta \gamma_{xz} \right] d\Omega \, dz - \int_{\Omega} q\delta \tag{5}$$

The governing equations of equilibrium can be derived easily from Equation (5) which is expressed as.

$$[C]\{\Delta\} = \{F\} \tag{6}$$

where $\{\Delta\} = \{U, V, W, X, Y, Z\}^T$ and $[C]$ is the stiffness matrix and $\{F\}$ is the force vector.

4 Numerical procedure

The thermomechanical bending analysis is conducted for combinations of metal and ceramic. For simplicity, Poisson's ratio of the two materials is assigned the same value.

Table1 contains the dimensionless center deflection $\overline{w}$ for a FG sandwich plate subjected to thermo-mechanical loads. The deflections are considered for $k = 3$ and different FG plate types. It can be seen from the table1 that the results of the present theory are very close to those of the other shear deformation theories. It can be observed that the HSDP overestimates the deflections comparatively to RHSDT and this, is due to the thickness stretching effect. For a sandwich plate, the deflections decrease when the dimensionless ratio (a/b) increases.
Table 2 shows the influence of power law index k, and different plate types on the transverse shear stresses of non-symmetric sandwich plates. As is observed, the increase of the power law index k leads to a decrease of stresses of the non-symmetric sandwich plates, and the thickness stretching effect leads also to a reduction of stresses.

Table1. Effect of aspect ratio a/b on the dimensionless deflection of the FGM sandwich plates (k=3)

Schem	Theory	a/b =1	a/b =2	a/b =3	a/b =4
(0-1-0)	RHDPT	**0.991418**	**0.386195**	**0.189408**	**0.109104**
	HDPT	1.04333	0.410224	0.203668	0.119285
	SSDPT	1.01673	0.399467	0.198203	0.115999
	TSDPT	1.03131	0.405363	0.201199	0.117801
	FSDPT	1.14369	0.450781	0.224252	0.131638
(1-3-1)	RHDPT	**0.962845**	**0.376115**	**0.184793**	**0.106656**
	HDPT	0.997264	0.392717	0.195100	0.114330
	SSDPT	0.971882	0.382438	0.189876	0.111191
	TSDPT	0.985791	0.388073	0.192741	0.112913
	FSDPT	1.09287	0.431354	0.214671	0.126040
(2-1-2)	RHDPT	**0.959755**	**0.375429**	**0.184632**	**0.106685**
	HDPT	0.990245	0.390314	0.194009	0.113755
	SSDPT	0.965082	0.380135	0.188853	0.110670
	TSDPT	0.978873	0.385715	0.191680	0.112363
	FSDPT	1.08505	0.428589	0.213340	0.125275
(1-0-1)	RHDPT	**0.959500**	**0.375360**	**0.184615**	**0.106689**
	HDPT	0.989882	0.390189	0.193956	0.113731
	SSDPT	0.964718	0.380017	0.188807	0.110654
	TSDPT	0.978509	0.385593	0.191630	0.112341
	FSDPT	1.08468	0.428454	0.213275	0.125235

Table2. Dimensionless transverse shear stresses τ_{xz} of the different sandwich square plates

		τ_{xz}		
k	Theory	**(0-1-0)**	**(1-3-1)**	**(2-1-2)**
0	RHDPT	**0.142430**	**0.205538**	**0.170443**
	HDPT	0.175814	0.227649	0.205372
	SSDPT	0.1716038	0.235615	0.198813
	TSDPT	0.1744812	0.231653	0.203034
	FSDPT	0.1736235	0.191050	0.212861
1	RHDPT	**0.136266**	**0.140042**	**0.140973**
	HDPT	0.169464	0.171405	0.171788
	SSDPT	0.163779	0.165731	0.166155
	TSDPT	0.167460	0.169410	0.169807
	FSDPT	0.174786	0.175641	0.176173
2	RHDPT	**0.114226**	**0.122363**	**0.124937**
	HDPT	0.149472	0.153330	0.152449
	SSDPT	0.140626	0.146338	0.147845
	TSDPT	0.146059	0.150751	0.150873
	FSDPT	0.165933	0.160245	0.154643
3	RHDPT	**0.104469**	**0.114044**	**0.116725**
	HDPT	0.140350	0.144499	0.142463
	SSDPT	0.130468	0.137199	0.138410
	TSDPT	0.136471	0.141781	0.141103
	FSDPT	0.160582	0.151749	0.143556
4	RHDPT	**0.100046**	**0.109840**	**0.112740**
	HDPT	0.136337	0.140173	0.137604
	SSDPT	0.126029	0.132632	0.133851
	TSDPT	0.132264	0.137348	0.136362
	FSDPT	0.158488	0.147725	0.138061
5	RHDPT	**0.0981636**	**0.107666**	**0.110923**
	HYDPT	0.134806	0.138122	0.135375
	SSDPT	0.124290	0.130319	0.131795
	TSDPT	0.130635	0.135185	0.134204
	FSDPT	0.158235	0.146089	0.135419

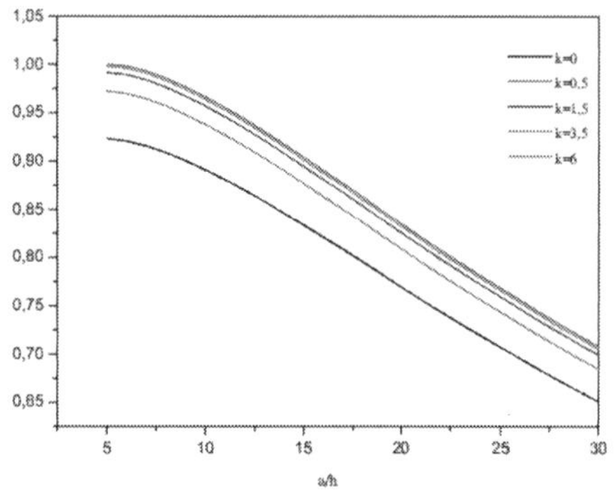

Fig.2 Dimensionless center deflection $\bar{w}$ as a function of side-to-thickness ratio a/h for sandwich plate type (1-3-1)

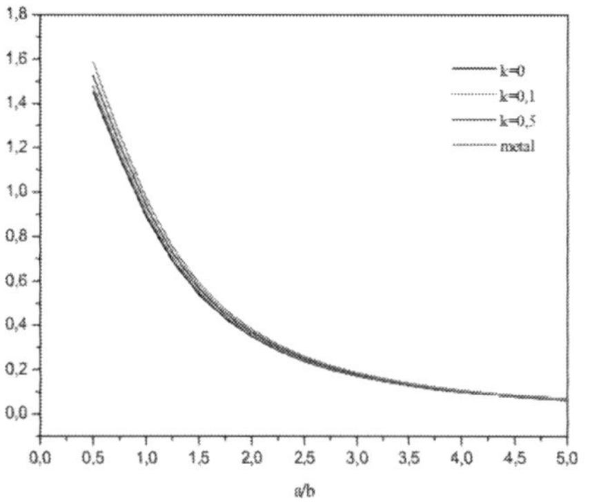

Fig.3 : Effect of the aspect ratio a/b on dimensionless center deflection $\bar{w}$ for sandwich plate type (1-3-1)

Figure 2 shows the variation of the center deflection $\bar{w}$ with side-to-thickness ratio a/h of sandwich plate type (1-3-1) with different volume fraction exponent k. The deflections of the sandwich plates decrease as a/h increases. Figure 3 shows the variation of the center deflection $\bar{w}$ with the aspect ratio a/b of sandwich plate type (1-3-1). The deflection of the ceramic plate is found to have the smallest magnitude and that of the metallic plate the largest one. The increase of the aspect ratio a/b leads to a decrease of deflections of the homogeneous and FG sandwich plates.

Conclusion

A novel hyperbolic shear deformation theory for thermo-mechanical bending of non-symmetric sandwich plates is presented. The theory accounts for the stretching and shear deformation effects without requiring a shear correction factor. The gradation of properties through the thickness is assumed to be of the power law type and comparisons have been made with homogeneous isotropic plates. The governing equations and boundary conditions are derived by employing the principle of virtual work. These equations are then solved via Navier-type solution. Numerical results for stresses and deflection are obtained and investigated for different plate configurations. The effect of power-law index, span to thickness ratio and aspect ratio are studied. It has been confirmed that the inclusion of thickness stretching effect makes the plate stiffer, and hence, leads to a reduction of deflections and stresses.

References

[1] M. Koizumi, FGM activities in Japan, *Composites Part B*, **28**, 1997, 1–4.
[2] J. N. Reddy, Analysis of functionally graded plates, *Int. J. Num. Meth. Eng*, **47**,2000, 663-684.
[3] Z.Q. Cheng, R.C. Batra, Deflection relationships between the homogeneous Kirchhoff plate theory and different functionally graded plate theories, *Arch. Mech*, **52**, 2000, 143-58.
[4] Z.Q. Cheng and R.C. Batra, Exact correspondence between eigenvalues of membranes and functionally graded simply supported polygonal plates, *J. Sound Vibration*, **229**, 2000, 879-895.
[5] C.T. Loy, K.Y. Lam, J.N. Reddy, Vibration of functionally graded cylindrical shells, *Int. J. Mech. Sci*, **41**, 1999, 309-324.
[6] A.M. Zenkour, Generalized shear deformation theory for bending analysis of function alley graded materials, *Appl. Math. Model*, **30**, 2006, 67-84.
[7] J.T. Katsikadelis, A.E. Armenakas, Plates on elastic foundation by BIE method, *J. Eng. Mech.*, **110**, 1984, 1086-1105.
[8] A. El Zafrany, S. Fadhil, K. Al-Hosani, A new fundamental solution for boundary element analysis of this plates on Winkler foundation, *Int. J. Numer. Meth. Eng*, **38**, 1995, 887-903.
[9] P.L. Pasternak, New Calculation Method for Flexible Substructures on a Two-Parameter Elastic Foundation, Gos. Izdat. Literatury po Stroitelstvu i Arkhitekture, Moscow, 1954, 1-56.
[10] S.S. Akavci, Buckling and free vibration analysis of symmetric and antisymmetric laminated composite plates on an elastic foundation, *J. Reinf. Plas. Comp*, **26**, 2007,1907-1919
[11] H. Saidi, M.S.A. Houari, A. Tounsi, E.A. Adda Bedia, Thermo-mechanical bending response with stretching effect of functionally graded sandwich plates using a novel shear deformation theory, *Steel and Composite Structures*, **15**, 221–245, 2013.

A MESHFREE METHOD FOR HEAT EXPLOSION PROBLEMS WITH NATURAL CONVECTION IN INCLINED POROUS MEDIA

Abdoulhafar Halassi[1], Youssef Joundy[2], Loubna Salhi[3,a], and Ahmed Taik[2]

[1] Faculty of Sciences and Techniques, University of Comoros, Moroni, Comoros.
[2] Faculty of Science and Technology Mohammedia, University Hassan II Casablanca, Morocco.
[3] Mohammed VI Polytechnic University, Benguerir, Morocco.

Abstract. This paper investigates the interaction between natural convection and heat explosion in porous media. A meshless collocation method based on multiquadric radial basis functions has been applied to study the problem in an inclined two-dimensional porous media. The governing equations consist of coupling the Darcy equations in the Boussinesq approximation of low density variations to the heat equation with a nonlinear chemical source term. The numerical results obtained are in good agreement with some previous studies that consider the vertical direction. A complex behaviour of solutions is observed, including periodic and aperiodic oscillations. We have shown that a small inclination of the container stabilizes the reactive fluid and can prevent thermal explosion.

Keywords: Heat explosion, Natural convection, Inclined porous media, Frank-Kamenetskii parameter, Multiquadric radial basis functions method.

1 Introduction

Consider an enclosure filled with fluid subjected to a temperature difference on two opposite walls, while the rest of the walls are adiabatic. The fluid beside the hot wall will be heated and due to its decrease of density buoyancy will carry it upwards. An inverse phenomenon will occur along the cold wall causing the fluid to be colder and denser and thus it will travel downwards. This natural convection phenomenon appears frequently in nature and is present in many industrial applications. One key issue concerning the heat transfer process in fluids, is how efficient it is, if it varies depending on some critical parameters. In heat explosion domain, the development of natural convection can significantly alter the heat transfer from the reacting fluid to the environment and hence alter the balance between heat production and heat loss, which determines whether or not the system will explode. This work may be considered as a continuation of a series of investigations concerning the interaction between heat explosion and natural convection [1, 6, 7]. Most researches in this area, like [2, 4], are concerned with the case when the fluid enclosure is oriented vertically or horizontally. In the present paper, we consider the case when the heat transfer process is produced in an inclined rectangular enclosure that is completely filled with a fluid-saturated porous medium. This is a complex process which involves various critical parameters like the Rayleigh number, the enclosure inclination angle and the Frank-Kamenetskii parameter. A range of numerical simulations is conducted by using the multiquadric radial basis functions (MQ-RBF) collocation method.

The paper is organized as follows: The governing equations are presented in section 2. The stream function-vorticity formulation is described in section 3. A brief description of the numerical method used is given in section 4, and numerical results are presented in section 5.

2 Governing equations

The geometry considered is an inclined rectangular enclosure with an inclination angle α as shown in Figure 1.

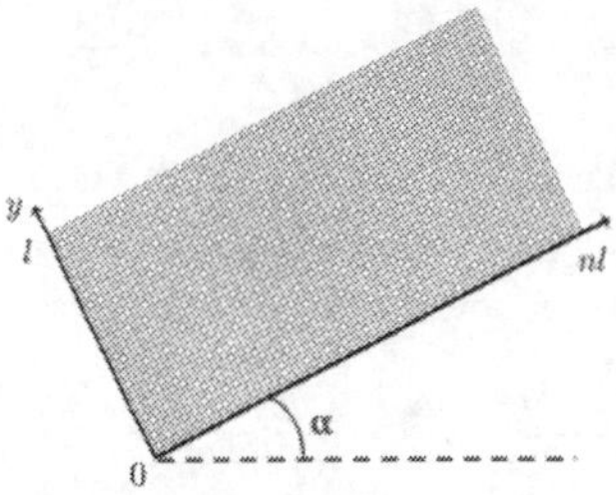

Fig. 1. Inclined fluid enclosure.

We consider the nonlinear heat equation coupled with Darcy equations for fluid motion in porous media. The fluid is Newtonian and all the thermophysical properties are supposed to be constant, except for the density in the buoyancy term that can be adequately modelled by the Boussinesq approximation, and that compression effects and viscous dissipation are neglected.

[a] e-mail: Loubna.SALHI@um6p.ma

Under these assumptions, the problem is described in a two-dimensional space, $0 \le x \le nl$, $0 \le y \le l$, by the following equations:

$$\frac{\partial T}{\partial t} + \mathbf{v} \bullet \nabla T = \kappa \Delta T + qK(T), \tag{1}$$

$$\frac{\partial \mathbf{v}}{\partial t} + \frac{\mu}{K_p}\mathbf{v} + \nabla p = g\beta\rho(T - T_0)\begin{pmatrix} -\sin\alpha \\ -\cos\alpha \end{pmatrix}, \tag{2}$$

$$\nabla \bullet \mathbf{v} = 0. \tag{3}$$

The boundary conditions will be specified below.
Here, T is the temperature, $\mathbf{v} = (u, v)$ the velocity, p the pressure, κ the coefficient of thermal diffusivity, q the adiabatic heat release, ρ the density, μ the coefficient of kinematic viscosity, g the gravity acceleration, β the coefficient of thermal expansion, T_0 is the characteristic value of the temperature and K_p is the permeability. The function $K(T)$ describes the reaction rate where the temperature dependence is given by the Arrhenius law

$$K(T) = k_0 \exp\Big(-\frac{E}{R_0 T}\Big),$$

where E is the activation energy supposed to be large in this problem, R_0 is the universal gas constant and k_0 is the pre-exponential factor.

3 Stream function vorticity formulation

In order to obtain the dimensionless form, we introduce the following spatial variables:

$$x' = x/l, \quad y' = y/l, \quad t' = \frac{\kappa}{l^2}t,$$

$$\mathbf{v}' = \frac{l}{\kappa}\mathbf{v}, \quad p' = \frac{K_p}{\mu\kappa}p \quad \text{and} \quad \theta = \frac{E(T - T_0)}{RT_0^2}.$$

For convenience, we keep the same notation for all variables except for the temperature, so we may re-write system (1)-(3) as follows:

$$\frac{\partial \theta}{\partial t} + u\frac{\partial \theta}{\partial x} + v\frac{\partial \theta}{\partial y} = \frac{\partial^2 \theta}{\partial x^2} + \frac{\partial^2 \theta}{\partial y^2} + F_K e^{\theta}, \tag{4}$$

$$\frac{\partial u}{\partial t} + u + \frac{\partial p}{\partial x} = -R_p \theta sin\alpha, \tag{5}$$

$$\frac{\partial v}{\partial t} + v + \frac{\partial p}{\partial y} = -R_p \theta cos\alpha, \tag{6}$$

$$\frac{\partial u}{\partial x} + \frac{\partial v}{\partial y} = 0. \tag{7}$$

With the boundary conditions:

$$x = 0, n \ : \frac{\partial \theta}{\partial x} = u = 0, \quad y = 0, 1 \ : \theta = v = 0.$$

For incompressible flows with constant fluid properties, Eqs. (4)-(7) can be simplified by introducing the stream function Ψ and vorticity ω as dependent variables.

We have $(u, v) = (\frac{\partial \Psi}{\partial y}, -\frac{\partial \Psi}{\partial x})$ and $\omega = \frac{\partial v}{\partial x} - \frac{\partial u}{\partial y}$.

We apply the rotational operator to Eqs. (5)-(6) in order to eliminate the pressure. We get

$$\frac{\partial \theta}{\partial t} + \frac{\partial \Psi}{\partial y}\frac{\partial \theta}{\partial x} - \frac{\partial \Psi}{\partial x}\frac{\partial \theta}{\partial y} = \frac{\partial^2 \theta}{\partial x^2} + \frac{\partial^2 \theta}{\partial y^2} + F_K e^{\theta}, \tag{8}$$

$$\sigma\frac{\partial \omega}{\partial t} + \omega = R_p\Big(\frac{\partial \theta}{\partial y}sin\alpha - \frac{\partial \theta}{\partial x}cos\alpha\Big), \tag{9}$$

$$\omega = -\Delta\Psi. \tag{10}$$

With the boundary conditions :

$$x = 0, n \ : \frac{\partial \theta}{\partial x} = \frac{\partial \Psi}{\partial y} = \Psi = 0,$$

$$y = 0, 1 \ : \theta = \frac{\partial \Psi}{\partial x} = \Psi = 0,$$

and some appropriate initial conditions on θ, ω and Ψ.

Here $F_K = \dfrac{Eqk_0\alpha e^{-\frac{E}{RT_0}} l^2}{RT_0^2\kappa}$ is the Frank-Kamenetskii parameter, $R_p = \dfrac{K_p\rho R_a}{l^2}$, where $R_a = \dfrac{g\beta RT_0^2 l^3}{E\kappa\mu}$ is the Rayleigh number. Under the assumptions of large activation energy, $RT_0/E \ll 1$, we can perform the Frank-Kamenetskii transform, so that the nonlinear reaction rate in Eq. (8) is taken to be $F_k exp(\theta)$ as shown in [3]. Next, $\sigma = \frac{1}{V_a}$ stands for the inverse of Vadasz number, $V_a = \dfrac{P_r}{D_a}$, $P_r = \dfrac{\mu}{\kappa}$ is the Prandtl number and $D_a = \dfrac{K_p}{l^2}$ is the Darcy number.

3.1 MQ-RBF method for numerical interpolation

To understand the effect of natural convection phenomena on heat transfer in the case of quasi-steady flow, we use the multiquadric radial basis function (MQ-RBF) method because of its popularity in many applications and its good approximation properties [5, 8]. The multiquadric is representive of the class of RBFs that are global, infinitely diffetentiable, and contains a shape parameter $\varepsilon \neq 0$ which plays an important role for the accuracy of the method. The MQ-RBF is radially symmetric about its center, $\mathbf{x} \in \mathbb{R}^2$, and its argument $r = ||\mathbf{x}||_2$ is dependent on the node location. It is defined by

$$\varphi(r, \varepsilon) = \sqrt{r^2 + \varepsilon^2}. \tag{11}$$

Now, let us assume that $(\theta^n, \omega^n, \Psi^n)$ are known at time $t = n\Delta t$. They can be approximated as follows:

$$\hat{\theta}^n \simeq \sum_{j=1}^{N} \alpha_j^n \varphi(\|\mathbf{x}_i - \mathbf{x}_j^c\|_2, \varepsilon), \quad \hat{\omega}^n \simeq \sum_{j=1}^{N} \beta_j^n \varphi(\|\mathbf{x}_i - \mathbf{x}_j^c\|_2, \varepsilon),$$

$$\hat{\Psi}^n \simeq \sum_{j=1}^{N} \gamma_j^n \varphi(\|\mathbf{x}_i - \mathbf{x}_j^c\|_2, \varepsilon),$$

where $\mathbf{x}_1^c, \mathbf{x}_2^c, \ldots, \mathbf{x}_N^c \in \Omega \subset R^2$ is a given set of centres with $\mathbf{x}_i^c = (x_i^c, y_i^c)$ and φ the MQ-RBF defined in (11).

By using the implicit scheme for the time derivative, Eqs. (8)-(10) become:

$$\theta^{n+1} - \Delta t(-\Psi_y^n \otimes \theta_x^{n+1} + \Psi_x^n \otimes \theta_y^{n+1} + \theta_{xx}^{n+1} + \theta_{yy}^{n+1}) = \theta^n + \Delta t F_K e^{\theta^n}, \quad (12)$$

$$\omega^{n+1} = \left(\frac{\sigma}{\sigma + \Delta t}\right)\omega^n + \left(\frac{\Delta t}{\sigma + \Delta t}\right) R_p\left(\theta_y^n sin\alpha - \theta_x^n cos\alpha\right), \quad (13)$$

$$\Delta \Psi^{n+1} = -\omega^{n+1}, \quad (14)$$

where $\otimes$ is the element-by-element matrix product.

4 Numerical simulations

In all our simulations, we fix the value of the Rayleigh number, $R_p = 10000$, the geometry of the domain, $n = 8$, $\sigma = 0.01$, $F_K = 3.8$, and we vary the angle of inclination α.

We start with the case where the reservoir is placed on an horizontal foundation, $\alpha = 0$ (no-inclination). Figure 2(a) shows the aperiodic behavior of the solution. Indeed, the system is aperiodic represented by a thick trajectory in the phase space of the mean of the temperature as a function of the mean of the stream function (see Figure 2(b)). A slight increase in the angle of inclination leads to a periodic regime. In Figure 3(a), we observe a periodic solution corresponding to the value of the inclination angle $\alpha = 0.01\pi$ as well as its phase space in Figure 3(b). If we continue to increase the angle α, the oscillations disappear and the solution converges to a steady state. Figure 4 shows the stationary solution for $\alpha = 0.011\pi$. In order to highlight the way in which the aperiodic solution appears or disappears by varying the angle of inclination of the container we have drawn the bifurcation diagram (see Figure 5). Two types of behavior can be identifed. In the first one, the attractor is represented by a set of points distributed over the vertical line ($\alpha \in [0; 0.0021\pi]$), in the second one, the attractor is represented by a finite number of points attained successively, which corresponds to a periodic or steady behavior. We remark that aperiod motion disappears when the angle α exceeds the value of 0.0021π. We observe a reverse period doubling cascade near $\alpha = 0.007\pi$. Figures 6(a) and 6(b) illustrate the solution obtained for the period 4 taking into consideration the inclination angle $\alpha = 0.00205\pi$.

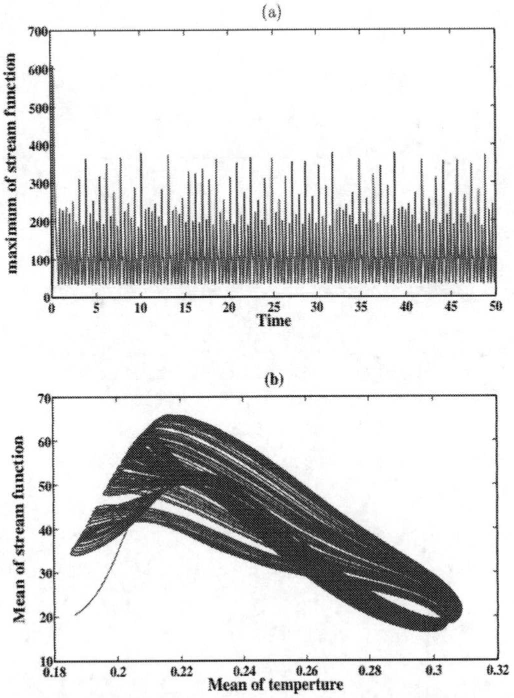

Fig. 2. (a) Maximum of the stream function against time and (b) mean of the stream function versus mean of the temperature for $R_p = 10000$, $\sigma = 0.01$, $n = 8$, $F_K = 3.4$ and $\alpha = 0$.

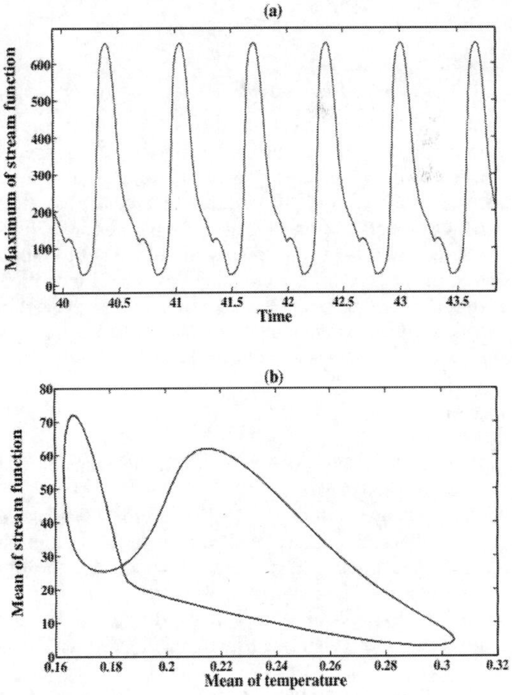

Fig. 3. (a) Maximum of the stream function against time and (b) mean of the stream function versus mean of the temperature for $R_p = 10000$, $\sigma = 0.01$, $n = 8$, $F_K = 3.4$ and $\alpha = 0.01\pi$.

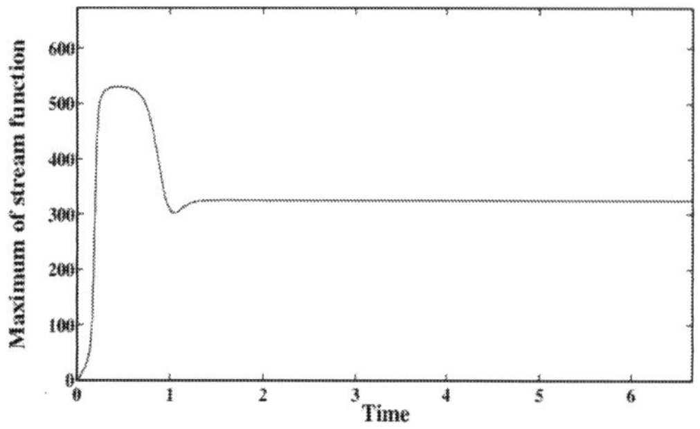

Fig. 4. Maximum of the stream function against time, for R_p = 10000, σ = 0.01, n = 8, F_K = 3.4 and α = 0.011π.

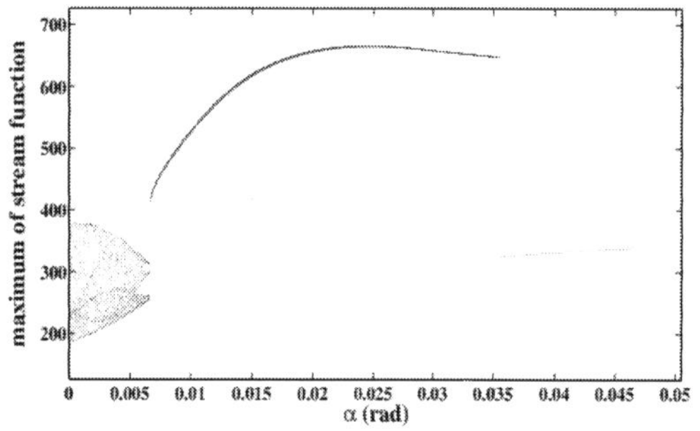

Fig. 5. Bifurcation diagram of the maximum of the stream function versus the inclinaison angle α with R_p = 10000, σ = 0.01, n = 8 and F_k = 3.8.

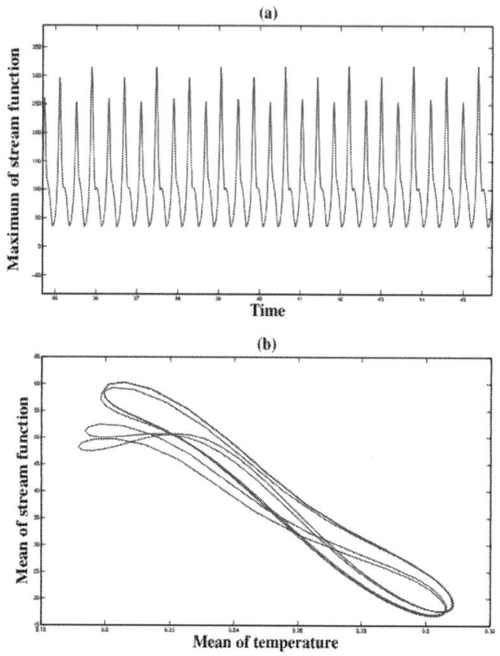

Fig. 6. (a) Maximum of the stream function against time and (b) mean of the stream function versus mean of the temperature for R_p = 10000, σ = 0.01, n = 8, F_K = 3.4 and α = 0.00205π.

5 Conclusion

In this work, the problem of heat explosion with natural convection is studied in an inclined porous media. The inclination angle is considered to be the control parameter. The MQ-RBF collocation method has been used for numerical simulations. In the case of an inclined medium, the instability decreases considerably by slightly increasing the angle of inclination. We found transitions between aperiodic and periodic oscillations through a sequence of period-doubling bifurcations.

References

[1] M. Belk, V. Volpert, Modeling of heat explosion with convection, Chaos **14**, 263-273 (2004).

[2] F. Bikany, A. Taik, K. Allali, V. Volpert, Numerical simulations of heat explosion with convection in porous media, Taylor and Francis, Combustion Science and Technology **187** (2014).

[3] D.A. Frank-Kamenetskii, Diffusion and Heat Transfer in Chemical Kinetics. Plenum Press, New York (1969).

[4] Y. Joundy, A. Taik, K. Allali, V. Volpert, Heat explosion in porous media using radial basis functions, MATEC Web of Conferences **83**, 07002 (2016).

[5] E. J. Kansa, Multiquadrics - A scattered data approximation scheme with applications to computational fluid dynamics. II. Solutions to hyperbolic, parabolic, and elliptic partial differential equations, Comput. Math. Appl. **19**, 8-9 (1990).

[6] A. Lazarovici, V. Volpert, J.H. Merkin, Steady states, oscillations and heat explosion in a combustion problem with convection, Euro. J. Mech. B. Fluids **24**, 189 (2005).

[7] A.S. Merzhanov, E.A. Shtessel, Free convection and thermal explosion in reactive systems, Acta Astronaut. **18**, 191 (1973).

[8] S.A. Sarra, J.E. Kansa, A Multiquadric Radial Basis Function Approximation Methods for the Numerical Solution of Partial Differential Equations, Marshall University and University of California, Davis (2009).

On the hysteretic operators with random parameters

Mikhail E. Semenov[1,2,3,4,*], *Olesia I.* Kanishcheva[2], *Peter A.* Meleshenko[1,2,**], *Olga O.* Reshetova[1], *Roman E.* Pervezentzev[2], and *Alexander V.* Budanov[5]

[1]Voronezh State University, Universitetskaya sq. 1, 394006, Voronezh, Russia
[2]Zhukovsky–Gagarin Air Force Academy, Starykh Bolshevikov st. 54 "A", 394064, Voronezh, Russia
[3]Voronezh State Technical University, XX-letiya Oktyabrya st. 84, 394006 Voronezh, Russia
[4]Geophysical Survey of Russia Academy of Sciences, Lenina av. 189, 249035 Obninsk, Russia
[5]Voronezh State University of Engineering Technology, Revolution av. 19, 394036 Voronezh, Russia

Abstract. In this work we introduce the novel class of hysteretic operators with random parameters. We consider the definition of these operators in terms of the "input-output" relations, namely: for all permissible continuous inputs corresponds the output in the form of stochastic Markovian process. The properties of such operators are also considered and discussed on the example of a non-ideal relay with random parameters. Application of hysteretic operators with stochastic parameters is demonstrated on the example of simple oscillating system and the results of numerical simulations are presented.

1 Introduction

Hysteresis phenomena are of interest both from the fundamental and applied points of view. Such an interest is caused by high incidence of these phenomena in a various technical systems (such as robotic, mechanical, electromechanical systems, management systems for tracking of aircrafts etc.) Also, these phenomena determine some unusual elasto-plastic properties of modern nanomaterials (such a properties are served as a base for construction of modern self-healing materials) on the basis of fullerene films. Moreover, the hysteretic phenomena are widely known in biology, chemistry, economics, etc. It should be noted that the hysteretic behavior of such systems is caused by either their internal structure, or the presence of separate blocks with hysteretic characteristics. Of course, when modeling the dynamics of such systems, it is necessary to use an adequate mathematical apparatus.

Currently used models of hysteretic phenomena both constructive (such as non-ideal relay, Preisach and Ishlinskii-Prandtl models, etc. [1]), and phenomenological (Bouc-Wen model, Duhem model, etc. [2]) assume a priori the stability of the parameters that identify the hysteretic properties of the corresponding operators. However, the stability of parameters in real engineering systems (e.g., in the systems modelled by the coupled inverted pendula [3]) does not always take place. In this way, such operators are the natural model in the situation when the parameters of hysteresis carrier are under influence of stochastic, uncontrollable affections. For example, it is difficult to control the switching numbers of non-ideal relay, which is a part of control systems of the corresponding devices, when the various external factors take place (in this case the switching numbers may be subjected to random changes). These circumstances make it necessary to develop the extended models of hysteretic effects, taking into account the stochastic changes of the parameters that are deteremined the corresponding hysteretic operators. We note that the equations with random parameters (principally such equations are linear) were considered in [4–6]. The strongly nonlinear differential equations containing the operator nonlinearity with random properties have not been considered in the literature. Thus, construction and investigation of the properties of hysteretic operators with random parameters seems novel and promising problem.

2 Non-ideal relay with random parameters

Let us consider a non-ideal relay (a detailed description of this and other hysteretic converters, as well as their properties in the case of deterministic parameters, are given in the classical book of M.A. Krasnosel'skii and A.V. Pokrovskii [7]), in which the switching numbers are not fixed, but are treated as random variables with absolutely continuous distribution function. Concerning these random variables, we make the following assumption: the probability density of each of the switching numbers will be assumed to be finite with non-intersecting supports. We denote these switching numbers as $\varphi_\alpha(u)$ and $\varphi_\beta(u)$. We will consider the case when the supports of the function $\varphi_\alpha(u)$ and $\varphi_\beta(u)$ are contained in the intervals $[u_\alpha^-, u_\alpha^+]$ and $[u_\beta^-, u_\beta^+]$, respectively.

Following the basic ideas presented in [7], the dynamics of the input-output relations for the operator of a non-ideal relay with random switching numbers is determined by two relations, namely: "input-state", "state-output".

*e-mail: mkl150@mail.ru
**e-mail: melechp@yandex.ru

We assume that all permissible for the converter R continuous inputs are given on the non-negative semi-axis $(t > 0)$ (the input-output relation for this converter has the form $x(t) = R[t_0, x_0, \alpha, \beta]\, u(t)$, $(t \geq t_0)$). The space of possible states of such an operator is defined as $\Omega = \Omega(\omega, p, u)$, $(\omega = 0, 1, 0 \leq p \leq 1, -\infty < u < +\infty)$.

The variable state of the converter $R\left[(1; p_0); x_0; \varphi_\alpha(u); \varphi_\beta(u)\right] u(t)$ is a random value that takes the value 0 with probability $(1 - p(t))$ and a value of 1 with probability $p(t)$. In other words it can be presented as a pair $\{1; p(t)\}$ (here the second output component corresponds to the probability that at the time t the first component is 1). The output of this converter is a random function $x(t)$ (Markovian process) taking a value of 1 with probability $p(t)$. The rule that determines the value of probability $p(t)$ will be given below.

2.1 Definition of input-output relation for a non-ideal relay with random switching numbers

Following the classical scheme proposed in [7], we give the definition of the input-output relation by means of a three-step construction:

- At the first step we define the input-output relation on the monotonic inputs only;
- At the second step, using the semi-group identity, the input-output relation is defined for all piece-wise monotonic inputs;
- At the third step, using the special limit construction, the corresponding converter will be defined for all monotonic inputs.

We define the operator R on the monotonic inputs. Let us assume that at the initial time point t_0 (to simplify the calculations, we assume that $t_0 = 0$) the operator R is in the state $1; p_0; u_0 \in \Omega, (u(0) = u_0)$. Let the input $u(t)$ be a monotonic increase, then for the time $t > 0$ the output is $x(t) = \{1; p(t)\}$ where

$$p(t) = \max\left\{p_0; \int_{-\infty}^{u(t)} \varphi_\beta(u)du\right\}. \tag{1}$$

The semi-group identity for the operator R immediately follows from the definition. Let t_1 be an arbitrary moment of time satisfying the inequality $0 < t_1 < t$, then the semi-group identity for the operator of a non-ideal relay has the form:

$$\begin{aligned} &R\left[t_0; p_0; u_0; \varphi_\alpha; \varphi_\beta\right] u(t) = \\ &= R\left[t_1; R\left[t_0; p_0; u_0; \varphi_\alpha; \varphi_\beta\right] u(t_1); u(t_1); \varphi_\alpha; \varphi_\beta\right] u(t). \end{aligned} \tag{2}$$

To define an operator on the piece-wise monotonic inputs (in the case of a finite interval $[0, T]$), we break this interval by points $t_1, t_2, \ldots, t_n$ in to intervals of monotonicity. On each of them we define the corresponding operator as an operator on a strictly monotonic input whose initial state will be defined as the state at the instant corresponding to the "last" change in the behavior of the input.

To determine the operator R on continuous inputs, we use the following limit construction. Let $u(t)(t \in [0, T])$ be an arbitrary continuous input. Let us consider an arbitrary sequence of piece-wise monotonic inputs $u_n(t)$, $(n = 1, 2, \ldots)$ that converges uniformly to each element of this sequence $u(t)$. A single-variable state $p_n(t)$, $(n = 1, 2, \ldots)$ will form a sequence of state variables $p_n(t)$, $(n = 1, 2, \ldots)$. Let us prove that the sequence $p_n(t)$, $(n = 1, 2, \ldots)$ converges uniformly. We estimate the absolute value of the difference:

$$|p_n(t) - p_m(t)| \leq \max_t \left|\int_{-\infty}^{u_n(t)} \varphi_\alpha(u)du - \int_{-\infty}^{u_m(t)} \varphi_\alpha(u)du\right| =$$

$$= \max_t \left|\int_{u_n(t}^{u_m(t)} \varphi_\alpha(u)du\right|. \tag{3}$$

Since the function $\varphi_\alpha(u)$ is continuous, and because of uniform convergence also

$$\lim_{n,m\to\infty} \max_t |u_n(t) - u_m(t)| = 0,$$

as well as, using the mean value theorem:

$$\max_t \left|\int_{u_n(t}^{u_m(t)} \varphi_\alpha(u)du\right| \leq \max_t \varphi_\alpha(t)\left[u_n(t) - u_m(t)\right]$$

the right-hand side of the inequality (3) tends to zero. Thus, the sequence of probabilities $p_n(t)$ is fundamental (the continuity is obvious), then there is $\lim_{n\to\infty} p_n(t) = p(t)$, which is comparable to an arbitrary continuous input $u(t)$.

2.2 Monotonicity of a non-ideal relay with random parameters

Let us consider the monotonicity property for the constructed converter. We determine the monotonicity with respect to the initial state of the non-ideal relay: if $\{u(t_0, x_0\}, \{v(t_0, y_0\} \in \Omega(\alpha, \beta)$, $x_0 \leq y_0$ and $u(t) \leq v(t)$ $(t \geq t_0)$, then $R[t_0, x_0, \alpha, \beta]\, u(t) \leq R[t_0, x_0, \alpha, \beta]\, v(t)$ $(t \geq t_0)$.

Let us note that this property can be used as the definition of a non-ideal relay. In order to use it, we define the outputs corresponding to monotonic inputs. Applying a semi-group identity, we define the outputs for piece-wise monotonic inputs. Further, this relation is extended by means of the special limit construction to all monotonic inputs. In this case, this relation will be the exclusive. We can also note a natural monotonicity in the switching numbers. With respect to the modified operator of a non-ideal relay with random parameters, the analogue of monotonicity can be presented in the form of the following theorem.

Theorem 1. Let $p\{x_{01} = 1\} \geq p\{x_{02} = 1\}$ and $x_1(t) \geq x_2(t)$. Then for any t: $p\{x_1 = 1\} \geq p\{x_2 = 1\}$.

3 Dynamics of a system under non-ideal relay with random parameters

In order to show the action of the developed operator on the real physical system let us consider the simple oscillating system under hysteretic force with random parameters.

Such a simple system is considered in [8, 9] and the external force has the form of a non-ideal relay with inversion of the switching numbers. One of the main results of these studies is the existence of unlimited solutions, namely, if the initial conditions are such that the hysteretic operator is triggered at the first cycle of oscillations, then the corresponding solution is unlimited, while the growth rate of the amplitude is proportional to the square root of time. Let us consider an analogous system with a non-ideal relay with switching numbers distributed according to an even-dimensional law. The equation of motion together with the corresponding initial conditions has the following form:

$$\begin{aligned} \ddot{x}(t) + \omega^2 x(t) &= R[t_0; p_0; u_0; \varphi_\alpha; \varphi_\beta]x(t), \\ x(0) = x_0, \ \dot{x}(0) &= x_1. \end{aligned} \tag{4}$$

To implement the numerical solution of the system (4), it is necessary to generate a set of random values corresponding to the switching numbers. Let us consider the case when $\varphi_\alpha(u)$ and $\varphi_\beta(u)$ corresponds to the uniform distribution law for α and β. For definiteness, we will assume that these functions correspond to uniform distributions in the intervals $[-1.5, -0.5]$ and $[0.5, 1.5]$, respectively. The solution of the system (4) can be obtained by specifying the initial conditions and the values of the switching numbers corresponding to the non-ideal relay. At each period, the switching numbers are selected from the corresponding distributions with the initial conditions for the next realization corresponding to the values of the phase coordinates obtained at the previous step. Using the described algorithm, a solution to the system (4) is obtained and the corresponding law of motion together with the phase portrait are shown in Fig. 1.

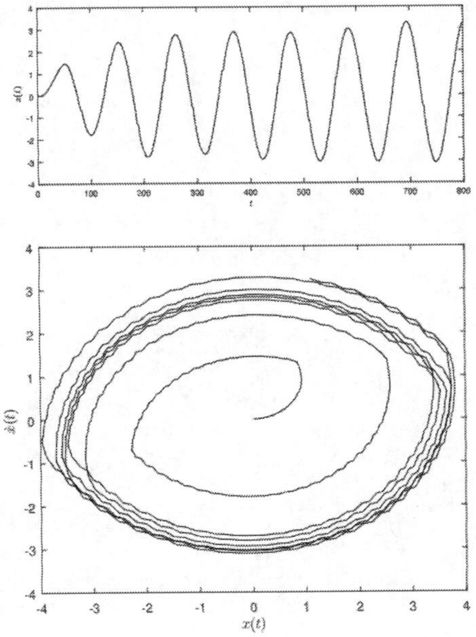

Figure 1. Solution (top panel) and phase portrait (bottom panel) for the system (4)

The following theorem characterizes the dynamics of the system (4).

Theorem 2. Let us suppose that the supports of the function φ_α and φ_β do not intersect. Then $\overline{\lim_{t\to\infty}} x(t) = \infty$, that is the amplitude tends to infinity with probability 1.

The proof of this theorem follows from the fact that the area of the minimal hysteretic loop is positive $S_{min} > 0$, as a consequence, the amplitude value at each cycle satisfies the inequality:

$$A_n^2(t) \geq nS_{min}$$

We also note that under the conditions of the theorem, the rate of growth of the amplitude with probability 1 is proportional to the square root of time.

4 Conclusions

The paper presents a generalization of the classical hysteretic converter in the form of non-ideal relay to the case when its switching numbers are randomly distributed according to a corresponding law. The properties of this converter are established (namely, the definition, together with the monotonicity), as well as the dynamics of the simple mechanical system in the form of oscillator under hysteretic force determined by a non-ideal relay with random parameters is considered.

This work is supported by the RFBR (Grants 16-08-00312-a, 17-01-00251-a, and 18-08-00053-a)

References

[1] M.E. Semenov et al., MATEC Web of Conference **83**, 01008(1)–01008(5) (2016)

[2] F. Ikhouane, V. Mañosa, J. Rodellar, Systems & Control Letters **56**, 197–205 (2007)

[3] M.E. Semenov et al., Archieve of Applied Mechanics **88**, 517–524 (2018)

[4] Øksendal Bernt, *Stochastic Differential Equations. An Introduction With Applications* (Berlin, Springer-Verlag, 2003) 379 p

[5] V.G. Zadorozhniy, Computational Mathematics and Mathematical Physics **53**, 486–502 (2013)

[6] V.G. Zadorozhniy, S.S. Khrebtova, Computational Mathematics and Mathematical Physics **49**, 1853–1868 (2009)

[7] Krasnosel'skii M.A., Pokrovskii A.V., *Systems with Hysteresis* (Berlin, Springer-Verlag, 1989) 410 p

[8] A.M. Solovyov et al., Procedia Engineering **201**, 549–555 (2017)

[9] A.M. Solovyov et al., Procedia Engineering **201**, 578–583 (2017)

Analysis of dynamics of a vertical cantilever in rotary coupling to the moving frame with movement limiters

Piotr Wolszczak[1,a], Grzegorz Litak[1,2,b], and Krystian Lygas[1]

[1] *Faculty of Mechanical Engineering, Lublin University of Technology, Nadbystrzycka 36, Lublin 20-618, Poland*
[2] *Department of Process Control, AGH University of Science and Technology, Mickiewicz Alley 30, Kraków 30-059, Poland*

Abstract. The efficiency of the mechanical energy harvesting with the use of vibrating elements can be improved by synchronizing stimulation vibrations and own linear frequencies of systems as well as super or sub harmonics induced by non-linear phenomena. The article presents numerical cross-sectional study of the mechanical system. The system consists of an elastic beam set vertically, which the lower end is fixed in the rotary support, and is stimulated to move in the horizontal axis. The upper end of the beam is free but below its level there are bumpers limiting the free rotation of the beam. Numerical studies took into account the variability of the frequency and amplitude of the excitation beam movement, and horizontal distance between bumpers. Beam deflection was observed, on the basis of which the amount of energy generated by the piezo element was estimated. Nonlinear phenomena and analysis of frequency synchronization of vibrations improving the energy effect of an energy generator are presented.

1 Introduction

In the field of energy harvesting, a mechanical systems using non-linearity are used to increase the efficiency of the system with the ambient vibration sources. The frequency and amplitude of the system excitation are directly affecting the power output of the harvesting device. Account for high energy density, deflecting piezoelectric layers were proposed [1] to transduce mechanical into electrical energy. The occurrence of the nonlinear phenomena are giving additional frequency broadness of the moderate power output via inclined nonlinear resonance characteristics, multiple natural frequencies as well as their sub-harmonics and super-harmonics or dynamical hysteresis occurring with frequency change [2–4]. For industrial applications, the frequencies from over few to few hundred Hertz are of interests [5–8]. In the present note, we present the results on the bistable vibrational energy harvester with additional amplitude limiters inducing impacts [9, 10]. Here we focus on energy efficiency with changing selected system parameters, namely: excitation parameters (amplitude and frequency), clearance between the angular amplitude limiters.

2 Energy harvesting system

The subject of the research was the elastic beam set in a vertical position (as in the system of the inverted pendulum) and mounted on a rotary handle. The rotating handle was placed on a trolley equipped with two stops to limit the rotation of the beam. An important function of the bumpers is to cause collisions and increase the deflection arrow of the elastic beam. The schematic picture of the mechanical resonator is shown in Fig. 1.

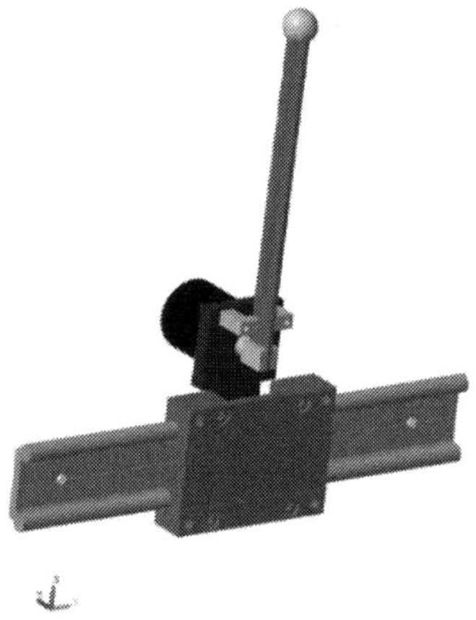

Figure 1. Schematic view of the resonator used in the Matlab program. The amplitude limiters (bumpers) are sketched in yellow. Piezoelectric layer can be placed above the contact point. Deflection of the beam will be proportional to the voltage output.

Sample time series of selected variables registered during the simulation is shown in Fig. 2. Note that at the initial conditions, the beam is in a vertical position, and then as a result of the movement of the trolley it falls onto

[a] e-mail: p.wolszczak@pollub.pl
[b] e-mail: g.litak@pollub.pl

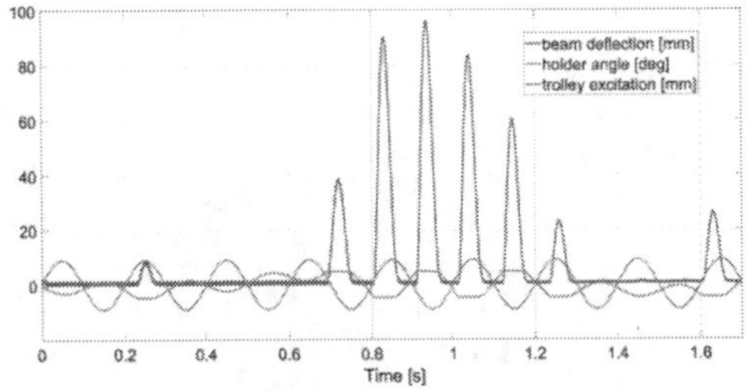

Figure 2. Sample part of the simulation results with sinusoidal excitation of the vertical beam lower end with frequency ω = 5Hz, amplitude A =9mm and distance between bumpers C = 10 mm (where the thickness of the beam is T =3 mm).

one of the bumpers (amplitude limiters). In the further part of the simulation, the beam is stimulated by the movements of the trolley for rotary motion and collisions with bumpers. The moment when the beam again reaches the vertical position and hits the opposite bumper is achieved for a series of selected parameters. In these cases, the kinetic energy of the rotating beam must be higher than the potential energy of the beam striving for a resting position, i.e. the support on one of the bumpers [12]. Note that for $C \to 0$ we get the inverted elastic pendulum case studied by Friswell et al. [13].

For this system the physical experiments and numerical simulations (in the Matlab program using additional Simmechanics libraries) were performed after [10, 12]. During the physical experiments carried out, the beam position and deflection was observed with the high-speed camera, [9] and a piezo element was placed on the beam's wall. The voltage was recorded in the circuit equipped with the resistor and next, the amount of generated electric energy was calculated [11, 14]. In this way, the energy efficiency of the system operation was evaluated, and the results obtained from numerical simulations were confirmed in physical experiment.

Consequently, a number of numerical simulations were performed to investigate the phenomenon of nonlinearity, evaluate the energy efficiency of the inverse beam system with buffers in the low frequency range of excitation of the beam motion.

During the simulation, the following variables were recorded: the position of the trolley, the deflection angle of the beam and the distance between the ends of the beam as a measure of beam deflection (assuming one-sided deflection, this value is proportional to arrow deflection).

A series of simulations were prepared for various sets of parameters, which were:
- amplitude of excitation of the beam in the range: A = 1 - 16 [mm],
- frequency of beam activation in the range ω = 1-16 [Hz],
- distance between buffers, gap in the range C = 0 - 11 [mm].

The maximum distances between the bumpers and the beam walls in the initial position were 5.5 mm, 11 mm in total.

3 Results

Fig. 3 presents the surfaces formed from the results of sum of beam deformation expressed by changing the distance between the ends of the beam [mm]) for different beam excitation frequencies. In the case of a zero distance between buffers and a beam in the charts there are minimum non-zero values. The beam then shakes and its deflection is low.

Figure 3 shows that the total deflection of the beam gets higher values as the clearance C increases. Generally, energy output increases with clearance, which presumably enable the beam to accelerate just before impacts. For frequency of 1 Hz of excitation the influence of the clearance and excitation amplitude is the simplest and can be identified by the simple proportional relation. Increasing the frequency we observe corrugation on the resulting energy surface. It can be noticed that a similar flattened shape (with respect to 1 Hz results) takes the surface for an excitation frequency of 16 Hz. This non-monotonous property against frequency focuses our range choice of ω. Additionally, It can be noticed that for a higher edge of the surface for the highest value of clearance increases with the excitation amplitude, unlike 1 Hz, where the decrease of the total beam deflection (and harvested energy output) with decreasing amplitude of the trolley for a clearance C of 11 mm has been observed.

To illustrate the nonlinear effect, which can be noticed as the larger output deviation by changing the bifurcation parameter, on the amount of energy output we calculated deviations from the reference surface. Here the set of results by 1 Hz excitation was chosen as the reference surface to all other results obtained in the experiment. The simulation results for this frequency allow to achieve high values of the total deflection of the beam. The highest values were observed in the case of 2 Hz frequencies. Consequently, Fig. 4 presents the negative differences in the results of the total beam deflection obtained for the trolley activation frequency equal to 1 Hz and other systems where the excitation frequency (trolley movement) was from 2 to 16 [Hz].

On the other hand, Fig. 5 indicate positive fragments of surfaces, not visible in the figure 4, where the differences of the total deflection of the beam to the reference surface for higher beam excitation frequencies were higher than zero.

4 Conclusions

We studied the energy harvesting system with the inverted elastic pendulum with amplitude limiters. Changing the excitation parameters (amplitude, frequency) and geometrical property of the system (clearance) we obtained systematic simulation results of the potential transduction efficiency. Interestingly, the clearance has a considerable effect for the whole range of the excitation frequency. It seems to be more important for the total energy output with respect to amplitude of excitation (Fig. 3). Amplitude of the excitation includes some fluctuation and the noncontinuous changes in the output energy which could

 https://doi.org/10.1051/matecconf/201824101021

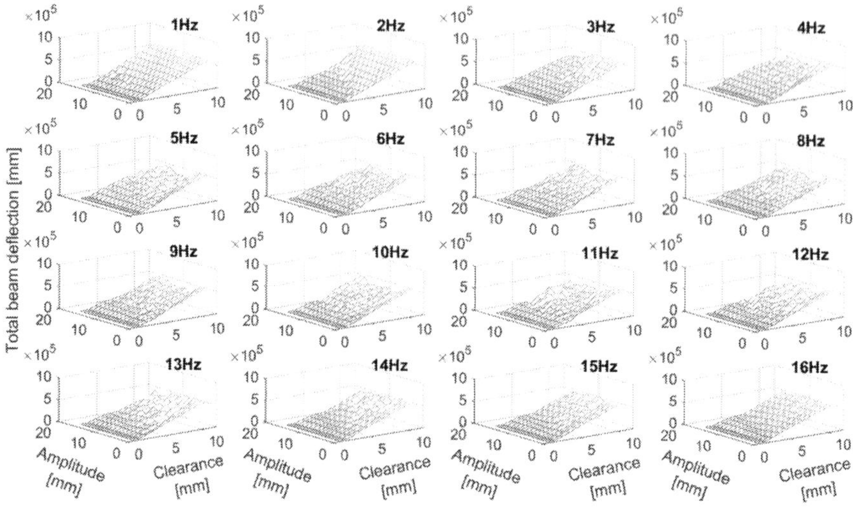

Figure 3. Surfaces formed from the results of strain distortion (sum of deflection [mm]) for different excitation frequencies and amplitudes (frequency from 1 to 16 [Hz] and amplitude from 1 to 16 [mm] trolley) and clearance between bumpers and beam in the range from 0 up to 11 [mm]. Z axis - sum of beam deflection [mm], X axis - clearance [mm], Y axis - amplitude [mm].

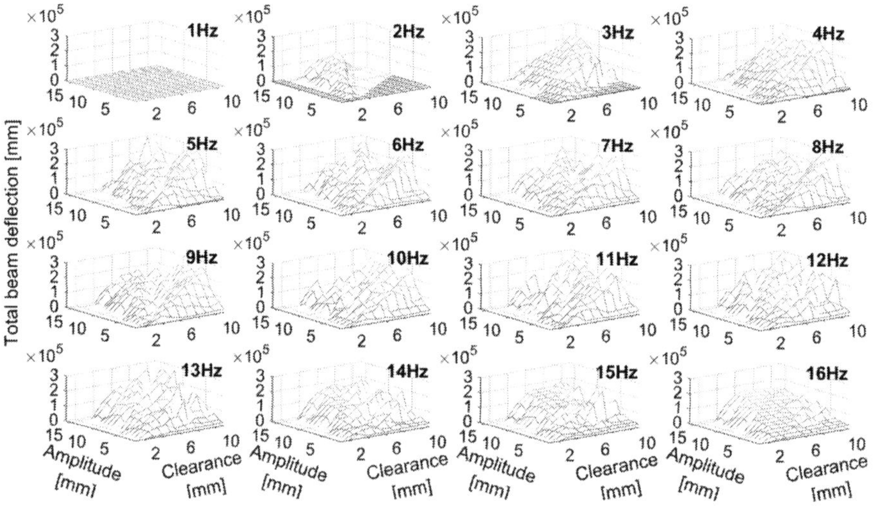

Figure 4. Differences between surfaces formed from the results of strain distortion (sum of deflection [mm]) for excitation frequencies 1 Hz and different excitation frequencies. Z axis - sum of beam deflection [mm], X axis - clearance [mm], Y axis - amplitude [mm].

reflect possible bifurcations. In the present studies we assumed only first mode solution. However, due to nonlinearities, the system was also sensitive to a super- and subharmonic resonance conditions. To complete these studies, more natural frequecies of beam oscillations should be investigated further with possibility of higher cantilever beam mode occurrence.

References

[1] S.P. Beeby, M.J. Tudor, N.M. White, Meas. Sci. Technol. **17**, R175 (2006)

[2] A. Syta, G. Litak, M.I. Friswell, S. Adhikari, Eur. Phys. J. B **89**, 99 (2016)

[3] T. Huguet, A. Badel, A., M. Lallart, Appl. Phys. Lett. **111**, 173905 (2017)

[4] S. Zhou, J. Cao, G. Litak, J. Lin, TM-Tech. Mess. https://doi.org/10.1515/teme-2017-0076 (2018)

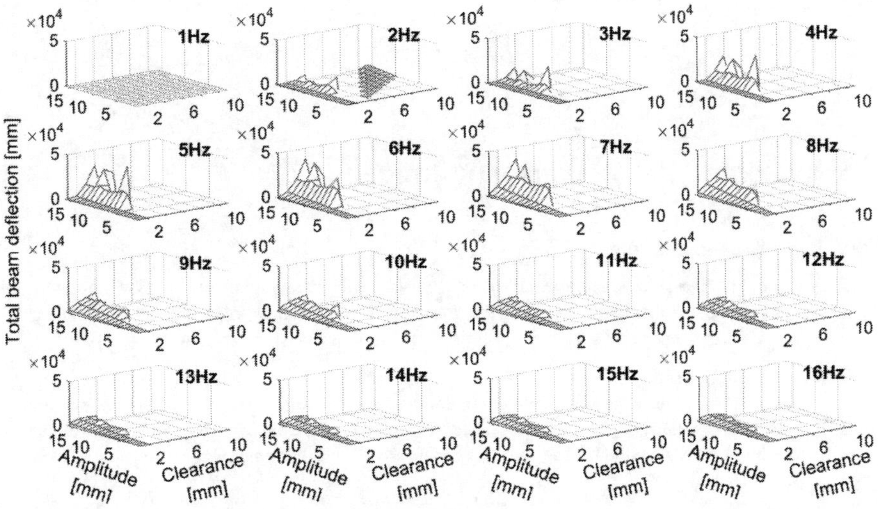

Figure 5. Fragments of surfaces of differences between results of strain distortion (sum of beam deflection [mm]) for excitation frequencies 1 Hz and different excitation frequencies where value were positive (Z axis - sum of beam deflection [mm], X axis - clearance [mm], Y axis - amplitude [mm].).

[5] P.D. Mitcheson, E.M. Yeatman, G.K. Rao, A.S. Holmes, T.C. Green, Proceedings of the IEEE **96**, 1457 (2008)

[6] R.L. Harne, K.W. Wang, Smart Mater. Struct. **22** 023001 (2013)

[7] S.P. Pellegrini, N. Tolou, M. Schenk, J.L. Herder, J. Intell. Mater. Syst. Struct. **24** 1303 (2013)

[8] J. Twiefel, H. Westermann, J. Intell. Mater. Syst. Struct. **24**, 1291 (2013)

[9] P. Wolszczak, K. Lygas, G. Litak, Mech. Syst. Sig. Process. **107**, 43 (2018)

[10] K. Lygas, P. Wolszczak, P. Staczek, G. Litak, in *Energy Harvesting for Wireless Sensor Networks: Technologies, Components and System Design* Eds. O. Kanoun and C. Viehweger (De Gruyter Oldenbourg, Berlin 2019)

[11] K. Lygas, P. Wolszczak, G. Litak, MATEC Web of Conferences **148** 16003 (2018)

[12] G. Litak, P. Wolszczak, K. Lygas, in *Vibration, Control and Stability of Dynamical Systems. DSTA 2017*, Eds. Awrejcewicz J. et al. (Department of Automation, Biomechanics and Mechatronics, ARSA, Lodz 2017) pp. 291-298.

[13] M.I. Friswell, S.F. Ali, S. Adhikari, A.W. Lees, O. Bilgen, G. Litak, J. Intell. Mater. Syst. Struct. 23, 1505 (2012)

[14] A. Koszewnik, P. Grzes, W. Walendziuk, Eur. Phys. J. Spec. Topics **224** 2719, (2015)

MATEC Web of Conferences **241**, 01022 (2018) https://doi.org/10.1051/matecconf/201824101022
CSNDD 2018

The control system of the yeast drying process

Piotr Wolszczak[1,a] and Waldemar Samociuk[2,b]

[1] *Lublin University of Technology, Faculty of Mechanical Engineering, Nadbystrzycka 36, Lublin 20-618, Poland*
[2] *University of Life Sciences in Lublin, Faculty of Production Engineering, 28 Głęboka Street, Lublin 20-612, Poland*

Abstract. The article presents the results of choosing how to control a real non-linear object. Yeast drying requires a precise temperature control due to the possibility of overheating. The object changes properties during of the process flow. Object identification is used and a mathematical model is developed. The model is used to select roboust control methods. The results are compared to the system of two PID regulators used in practice.

1 Introduction

When designing control systems, there is an inevitable compromise between the complexity of the controlled process model and the effort necessary to obtain a reliable mathematical description. The design methods can be broadly divided into two main classes: the indirect class (model-based approach) and the direct class (model-free approach) [1]. Finding models from the laws of physics can be difficult and time-consuming. Therefore control systems are designed based on experimental data. Indirect approaches use the collected data to identify a controlled process model. This method is based on the system identification tools [2–8]. The mathematical model (identifeid with data or derived from physical laws) is used to synthesize the controller in accordance with the design goals. On the other hand, the direct approach is not aimed at obtaining a precise model of the controlled process. In direct approaches, the identification step is limited to the basic process properties (such as static characteristics, time constant and delay). These properties are used to determine the structure of the control system (controller type) and the settings of this controller.

The reasons for the uncertainty of the models of object can be: using linearization to describe non-linear objects approximate or variable model parameters, errors of the identification based on measurement data, no description of parameters in the range of higher sampling rates, changing the object working conditions.

It is particularly difficult to design control systems for processes in which the model parameters change their values. In such situations, it must be taken into account the stability of the control system during these changes. In the literature for such systems there is a term of robust control which takes into account the object model (Plant) [9, 10]. The robust control system is designed so that the controller operates correctly in conditions when certain parameters of the object are uncertain (imprecisely determined or will change in a limited range).

Robust control requires a priori information about the range of uncertain or limits of properties variability and ensures that if the changes are within limits, the change of the adopted process control principle is no necessary.

1.1 Internal Model Control - Principle of operation

Internal Model Control (IMC) is a control that uses an internal process model. Assuming that the controller is an exact approximation of the real process, the output of the system will always be equal to the set value. However, designing a controller that is the exact opposite of the process is not always realizable. The use of a feedback loop is necessary if the process model is inaccurate or incomplete. A loopback system eliminates the effects of an inaccurate model to a certain extent and increases the system's robustness to interference and noise affecting the control object.

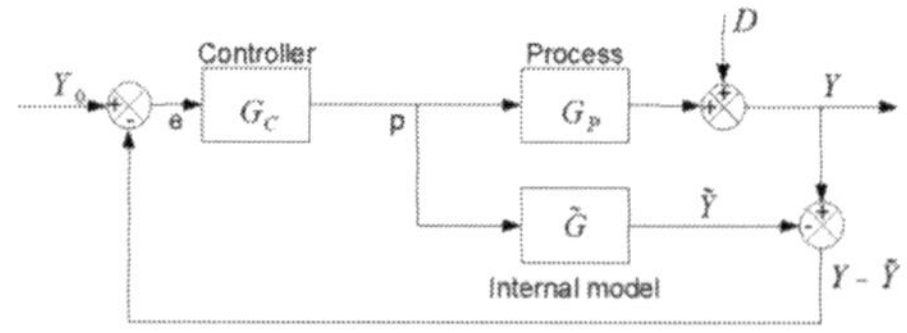

Figure 1. Internal Model Control (IMC) Diagram. Y_0 - sepoint signal, e - error signal, p - control signal, D - interference, $\hat{Y}$ - output of the internal model, Y - output signal.

[a] e-mail: p.wolszczak@pollub.pl
[b] e-mail: waldemar.samociuk@up.lublin.pl

2 Yeast drying

Drying yeast has an adverse effect on their fermentation capacity. The level of degradation depends on the method and parameters of the drying process [11]. During drying, cellular stress occurs [12]. Figure 2 presents offset of the drying surface during drying process of organic material. This offset has an effect on changing the surface of the drying surface. Additionally, lowering the surface of the material influences the change in moisture transport speed. The value of temperature and the rate of its change affect the viability of dried yeast [13, 14]. During industrial production are used various types (strains) of yeast, that have varying degrees of resistant to temperature [15, 16]. It is possible to use additives (skim milk) to immunize yeast for high temperature [17].

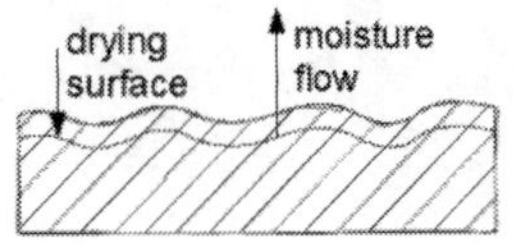

Figure 2. The offset of the drying surface during drying process of organic material.

Drying yeast is an industrial process, non-stationary and non-linear. For such processes the control is difficult. The selection of fixed parameters of the controller of the classical control system for the entire drying cycle is impossible, therefore it is necessary to use complex control systems [18]. For the sake of process safety, taking into account primarily material losses, it is advisable to use security systems and analyze the risk of failure [19, 20].

Figure 3 presents a scheme of a dryer equipped with one heater located outside the drying chamber, which operates without air recirculation. The air used for drying is mixed from two air streams. Thermodynamic transformations accompanying this process are shown in Figs. 4. The drying process in the adiabatic process is different than in the real case. The figure shows the heat loss index Q_{loss} to the dry air mass flow L. The value of Δi_{ad} means the difference between enthalpy of humid and atmospheric air.

3 Subject of research

The object of the research was the process of industrial yeast drying. The layer of wet yeast is placed on a sieve through which heated air flows. The air temperature is obtained by mixing streams of cool (constant temoerature) and heated air. The control signal is the degree of opening of the hot air throttle. The temperature of the yeast is monitored and controlled during the drying cycle. The drying cycle takes about 450 minutes. During the drying cycle, the temperature of the yeast layer is gradually raised. The yeast temperature control is carried out according to the defined drying curve shown in figure 5. Yeast temperature can not exceed 50 degrees, which would destroy them.

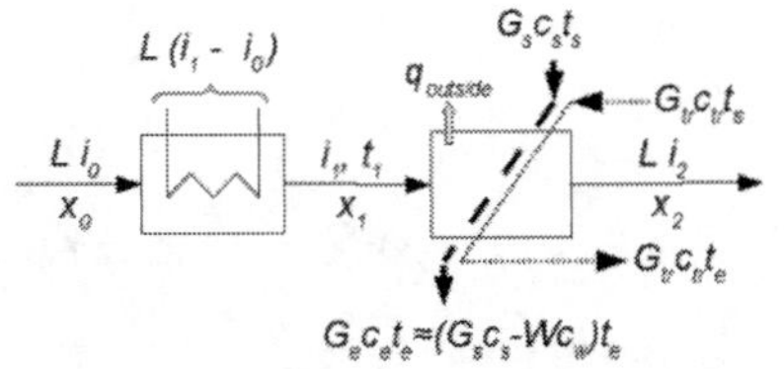

Figure 3. Block diagram of the continuous drying process (normal variant), a), b), c) and d) - heat streams flowing into the dryer, e), f), g) and h) - heat streams flowing out of the dryer; heat balance: a + b + c + d = e + f + g + h; left block - preheater, right block - drying chamber. L - mass flow of dry air [kg/s], G - mass flow of moist material [kg/s], c - specific heat [kJ/(kg*K)], i - specific enthalpy [KJ/kg] of dry air, x - absolute humidity of air [kg of steam/kg of dry air], t - temperature [deg C], W - mass flow of evaporated water from the material [kg/s]; indexes: s - initial value (start), w - water, e - finish value (end), tr - transporter, 0 - atmospheric air, 1 - heated air, 2 - humidified air. Based on [21]

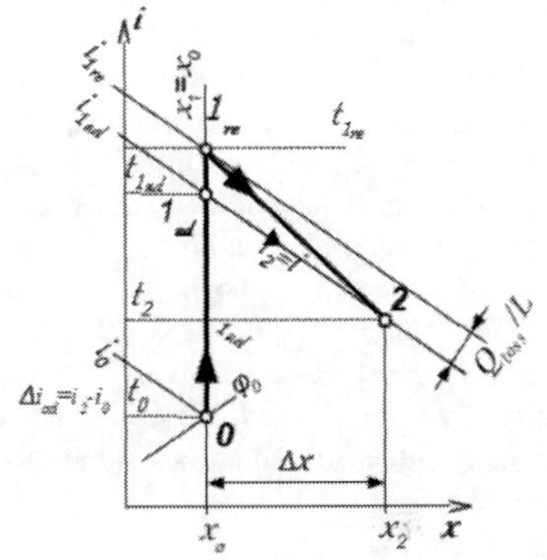

Figure 4. Simplified enthalpy diagram with thermodynamic curves of the drying air in a dryer (normal variant), $0 - 1_a d$ - heating in the adiabatic process, $0 - 1_r e$ - heating in the real process, $1_a d - 2$ - adiabatic dampening, $1_r e - 2$ - real dampening. i - specific enthalpy [KJ/kg] of dry air, x - absolute humidity of air [kg of steam/kg of dry air], t - temperature [deg C] ϕ_0 - relative humidity of atmospheric air; indexes: 0 - atmospheric air, 1 - heated air, 2 - humidified air, ad - adiabatic process (theoretic), re - real process (with losses). Based on [21]

Figure 5 presents ten series of drying cycles that contains the following signals: setpoint signal, throttle opening, temperature under the sieve (temperature of the mixture of cool and heated air) and temperature of the yeast layer. The figure shows a block diagram of a closed-loop control system containing the mentioned elements.

During the drying process a change in the physical properties of the object is observed. This change is revealed in the form of increasing the value of the error (the difference between the expected value and the adjusted value). Changing the physical properties of the object also involves changing the sensitivity of the object to changes in the control signal, which increases the risk of https://doi.org/10.1051/mateconf/201824101022

overheating the yeast.

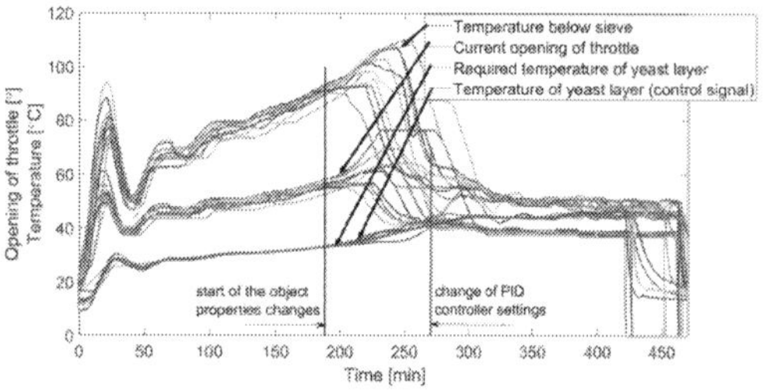

Figure 5. Example sets of signals aquisiited during drying process. Signals: setpoint trace - required temperature of yeast layer on sieve [°C], control signal - hot air throttle opening [°], the controlled variables - temperature below sieve and yeast layer temperature [°C].

A typical way to solve this problem is to change the controller's settings during the process. The control system defines the time of switching the controller's operation with the first set to the second set of parameters. One of the problems of such a solution is maladjustment to the moment of real change of the object's property and its character. Changing the settings of the regulator is introduced by step function, while the change in properties can be gradual. One of the ways to solve the problems could be the extension and smoothing of the drying curve. This solution does not ensure full security of the object, and additionally extends the duration of the process and increases its cost.

4 Methods and results

The procedure for developing control using the classical PID controlled and IMC system consisted of: registration of the signals from the real object, identification of the mathematical model of the object, (estimated by matching of the output signals) and preparing the systems simulation. During the identification of the model, the order and coefficients of the objects transform functions are adjusted. For every cycle (signals set) has been prepared a set of transform functions with different order value of characteristic equation. The estimation of nonlinear function was also conducted to characterize the object. Identification of the mathematical model was based on the output signals in the entire drying cycle (symbol AB) and separately for the first (A) and a second (B) parts. Moreover, due to the use of a temperature signal below the sieve, it was possible to identify the properties of the yeast layer.

Figure 5 presents a block diagram of a closed control system for the yeast drying process. The preparation of a block diagram of the process of drying yeast and the identification of mathematical models of system objects is also helpful in analysis and improving the currently operating control system. The model was used to optimize the previous PID controller settings and the time of switching the PID controller settings. Based on the observations of the raw signal curves in figure 5, it can be seen that fluctuations in the input and control signals occurs long before the switching of the PID controller settings. The figure 6 presents results of the system simulation in which the changeover time of the controller's settings was corrected. Such readjustment should protect the product from overheating and the controller output signal from saturation (what can be observed in figure 5). The object is descrbed by two transfer function G_A, G_B switched in 180th time unit of simulation, presented below with approximated values:

$$G_A = \frac{34s^3 + 2.5s^2 + 0.001234s}{s^4 + 33.4s^3 + 2.466s^2 + 0.002334s}, \qquad (1)$$

$$G_B = \frac{24.93s^3 + 0.1079s^2}{s^4 + 25.09s^3 + 0.1072s^2}. \qquad (2)$$

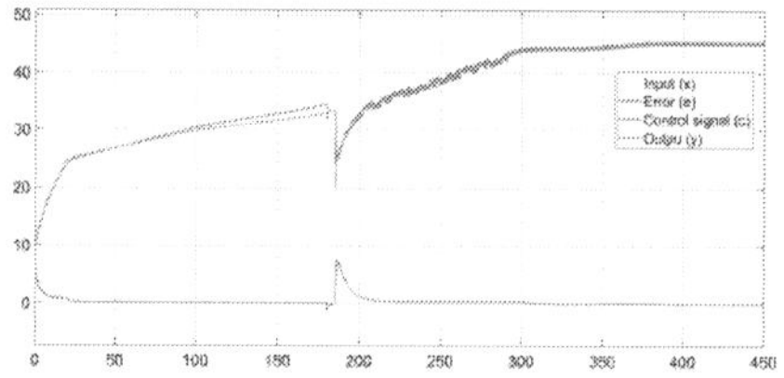

Figure 6. Simulation of drying process characterized by two transfer function G_A, G_B, and two controllers PI (Kp = 1.7, Ki = 0.785) and PID(Kp = 1.3, Ki = 0.312, Kd = 0.758), witch limited output in range from 20 to 80 [°] switched in 180th time unit of simulation.

Controlling the process changing its properties using one PID controller proved to be ineffective. To control the process of yeast drying throughout the cycle, two sets of parameters were prepared using a simple sliding-mode model. In order to improve the control quality and product protection from overheating, the robust method was used. The roustness controller was prepared using a mathematical model of the object, wherein the model was identified from signals recorded from the real object in whole drying cycle.The uncertainty of the characteristic equation coefficients in the range of +/- 20% of nominal values was used. Figure 7 presents a comparison of the step response of PI control systems 1 in the first part of the drying cycle (A), controller 2 (PID) controlling the process in the second part of the drying cycle (B) and the robust controller controlling the process in whole cycle (object transmittance identified on the basis of the signals in both parts of the process, parts A and B). However, the use of the controller 1 for controlling the part B and the controller No. 2 for controlling the model determined on the basis of part A causes instability of both systems. In figure 8 shows the results of the system sensitivity evaluation. Feedback of the inverted model was used

$$G_S = 1/G_{model}. \qquad (3)$$

The use of a robustness regulator allows the use of one set of settings throughout the entire drying cycle. In this way, fluctuations in the output signal caused by a sudden change of the object's properties and a step change in the control parameters were eliminated.

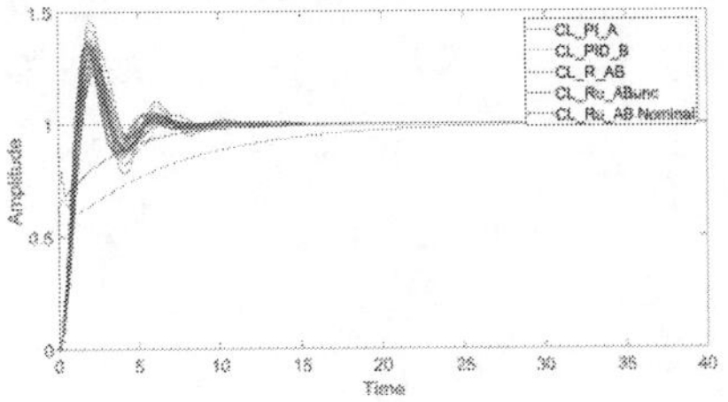

Figure 7. Step response for mathematical models of feedback systems consists of controllers and transfer function of objects: CL_PI_A - PI controller and transfer function of part A, CL_PID_B - PID controller and transfer function of part B, CL_R_AB - Robust controller and transfer function of parts A+B.

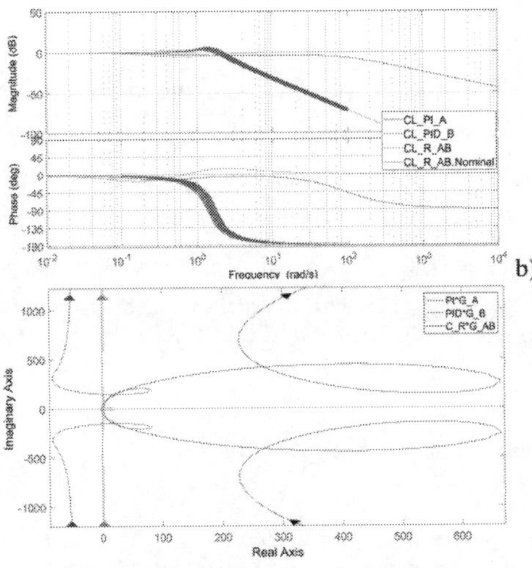

Figure 8. Sensitivity analysis a) Bode plots for closed-loop systems and b) Nyquist plot for mathematical models of open-loop systems consists of controllers and transfer function of objects: $PI * G_A$ - PI controller and transfer function of part A, $PID * G_B$ - PID controller and transfer function of part B, $C_R * G_A B$ - Robust controller and transfer function of parts A+B.

5 Conclusions

A change in the properties of yeasts during drying causes a non-linear response of the temperature signal. This change occurs so quickly that fluctuations in the yeast temperature signal pose a threat to them. The use of robustness controller eliminates this risk and improves quality control. A system with a robust controller reduces the time of adjustment, which is important in the case of sudden changes in the properties of the object. Use of robust controller for non-linear process control is known in the literature, but rarely used in practice. Therefore, it is advisable to test the robustness methodology in industrial conditions.

References

[1] Battistelli G. et. all., International Journal of Robust and Nonlinear Control **28**,3694-3712 (2017)

[2] Hjalmarsson H., Automatica **41**, 393-438 (2005)

[3] Wolszczak P., Łygas K., Litak G. Mechanical Systems and Signal Processing **107**, 43–52 (2018)

[4] Cechowicz R., Staczek P., Maintenance and reliability **16(2)**, 198-202 (2014)

[5] Martynenko A., Food Engineering Reviews **9** 91–111 (2017)

[6] Yang Li-Ping, Ding Shun-Liang, Litak Grzegorz, Song En-Zhe, Ma Xiu-Zhen, Chaos **25** 013105:1-13 (2015)

[7] Harris P., Arafa M., Litak G., Bowen C. R., Iwaniec J., European Physical Journal B **90**, 1-11 (2017)

[8] Litak G., Rusinek R., *4th IEEE International Conference on Nonlinear Science and Complexity (NSC)* (IEEE, Budapest, Hungary, 2012) 41-44

[9] Garcia C. E., Morari M. Industrial & Engineering Chemistry Process Design and Development **21**, 308-323 (1982)

[10] Garcia C. E.Morari M. Industrial & Engineering Chemistry Process Design and Development **24**, 472-484 1985

[11] Kamińska-Dwórznicka A., Skoniecka A., Advances of Agricultural Sciences Problem Issues (in polish) **573**, 35-42 (2013)

[12] Gamero-Sandemetrio E., Payá-Tormo L., Gómez-Pastor R., Aranda A., Matallana E., Microbial Cell (Graz, Austria) **5(4)**, 184-197 (2018)

[13] Gervais P., Maranon I., Biochemica et Biophysica Acta **1235, 1**, 52–56 (1995)

[14] Mensonides FI., Brul S. , Hellingwerf KJ. , Bakker BM. , Teixeira de Mattos MJ., The FEBS Journal **281(3)**, 825-841 (2014)

[15] Charoensopharat K., Thanonkeo P., Thanonkeo S., Yamada M., Antonie van Leeuwenhoek **108(1)**, 173-190 (2015)

[16] Techaparin A., Thanonkeo P., Klanrit P., Brazilian Journal of Microbiology **48(3)**, 461-475 (2017)

[17] Lee SB., Choi WS., Jo HJ., Yeo SH., Park HD., AMB Express **6(1)**, 105 (2016)

[18] Muhammad D., Ahmad Z. Aziz N., *Proc. of the 5th International Symposium on Design, Operation and Control of Chemical Processes* (PSE Asia, Singapore, 2010) 812-821

[19] Samociuk W. at all, Przemysł Chemiczny, **95(5)**, 1000 – 1003 (2016)

[20] Samociuk W., Wyciszkiewicz A., Gołacki K., Otto T., Przemysł Chemiczny, **96(8)**, 1763-1766 (2017)

[21] Ryniecki A., *Convection drying continuous: the exercise of design basics (in polish)* (University of Life Sciences, Poznan, Poland) 3-11

Reliability modeling and prediction of passive controlled structures through Random Forest

Weizhen. You[1], *Saidi* Alexandre[1], *Mohamed* Ichchou[1,*], *Zine* Abdel[1], and *Xiaopin* Zhong[2]

[1]LTDS, Ecole Centrale de Lyon, 69134 Ecully, France
[2]Shenzhen University, 518060 Shenzhen, China

Abstract. Reliability prediction plays a significant role in risk assessment of engineering structures. Mathematically, the prediction task can be seen as a classification (regression) procedure. In this aspect, machine learning methods have recently shown their superior performance over others in various research domains. Random forest (RF) is distinguished for its robustness and high accuracy in modeling and prediction work. However, its application in the area of structural reliability has not been widely explored. This study aims to explore the feasibility of RF as well as examine its performance in modeling and prediction of structure reliability in passive control mode. A numerical example is introduced in the simulation part to evaluate performance of the proposed method in different perspectives.

1 Introduction

The reliability of an engineering structure varies during its lifetime. To guarantee the desired level of its efficiency and safety, increasing efforts have been devoted to reliability prediction techniques. Traditional reliability models are generally based on specific a priori assumptions about lifetime distributions, which are no longer realistic for today's complex structures [1]. For this problem, growing attention has been paid to statistical learning approaches. Generally, the prediction task can be seen as a classification/ regression procedure. There are many machine learning models among which support vector machine (SVM) [2] and artificial neural network (ANN) [3] were most frequently considered in recent articles. However, the ANN is known for its complex architecture optimization, low robustness and enormous training time [4]. Besides, the SVM is time consuming for large-scale applications and sometimes shows large error in sensitivity calculations. Random forest (RF) is a tree method distinguished for its robustness and high accuracy in modeling and classifications [5]. However, RFs have rarely been used for reliability predictions in current literatures. Thus, this study focus on the feasibility of RF as well as its performance in reliability modeling and prediction of structures in passive control mode.

2 Structural response analysis

As is shown in Figure 1, a typical single degree-of-freedom (SDOF) system consists of a primary structure and a TMD which are characterized by the natural frequencies $\omega_T = \sqrt{k_T / m_T}$, $\omega_S = \sqrt{k_S / m_S}$, damping ratios $\xi_T = c_T / 2\sqrt{k_T m_T}$, $\xi_S = c_S / 2\sqrt{k_S m_S}$ and mass ratio $\mu = m_T / m_S$. The base acceleration $\ddot{y}_b$ which is modeled by Kanai-Tajimi stationary stochastic process, is expressed by $\ddot{y}_b = \ddot{y}_f + w(t) = -(2\xi_f \omega_f \dot{y}_f + \omega_f^2 y_f)$, where $w(t)$, ω_f , ξ_f , y_f are the stationary Gaussian white noise process, natural frequency, damping ratio and relative response of the elastic filter. The global state space vector of the structural system is constructed as $\mathbf{Z} = [y_T, y_S, y_f, \dot{y}_T, \dot{y}_S, \dot{y}_f]^T$, then the equation of motion is expressed in the state space form as

$$\dot{\mathbf{Z}} = \mathbf{A}\mathbf{Z} + \mathbf{f} \tag{1}$$

Here $\mathbf{f} = [0, 0, 0, 0, 0, -w(t)]^T$. $\mathbf{A} = [\mathbf{O}_{3\times3}, \mathbf{\Lambda}_{3\times3}; \mathbf{A}_L, \mathbf{A}_R]$ where

$$\mathbf{A}_L = \begin{bmatrix} -\omega_T^2 & \omega_T^2 & \omega_f^2 \\ \mu\omega_T^2 & -(\mu\omega_T^2 + \omega_S^2) & \omega_f^2 \\ 0 & 0 & -\omega_f^2 \end{bmatrix}, \quad \mathbf{A}_R = \begin{bmatrix} -2\xi_T\omega_T & 2\xi_T\omega_T & 2\xi_f\omega_f \\ 2\mu\xi_T\omega_T & -2(\mu\xi_T\omega_T + \xi_S\omega_S) & 2\xi_f\omega_f \\ 0 & 0 & -2\xi_f\omega_f \end{bmatrix}$$

The standard state space expression is

$$\begin{aligned} \dot{\mathbf{x}} &= \mathbf{A}\mathbf{x} + \mathbf{B}\mathbf{u} \\ \mathbf{y} &= \mathbf{C}\mathbf{x} + \mathbf{D}\mathbf{u} \end{aligned} \tag{2}$$

Here $\mathbf{x} = \mathbf{Z}$, $\mathbf{Bu} = \mathbf{f}$; $\mathbf{C} = [0,1,0,0,0,0]^T$; $\mathbf{D}$ equals to 0. The response process $\mathbf{y}$ can be simulated by softwares such as Matlab.

* Corresponding author: mohamed.ichchou@ec-lyon.fr

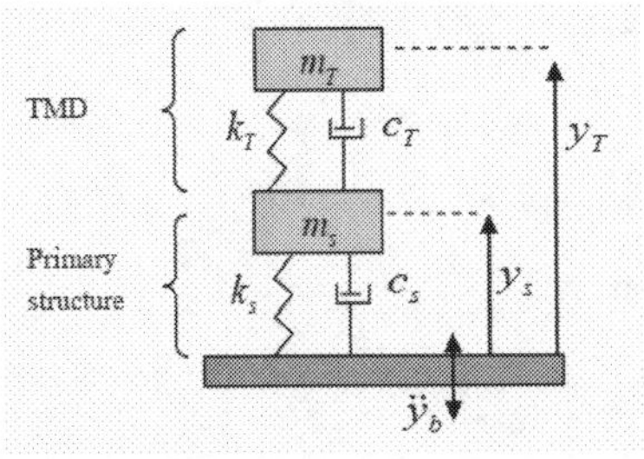

Figure 1 Basic structure of TMD system

3 The proposed method

The basic procedures of the proposed methodology is schematically described in Figure 2. We assume the uncertainties in the base structure are characterized by normal distributions. The structure failure probability in the proposed method is defined as the risk that the structure responses exceed a predefined limit within a certain interval of time [0, T].

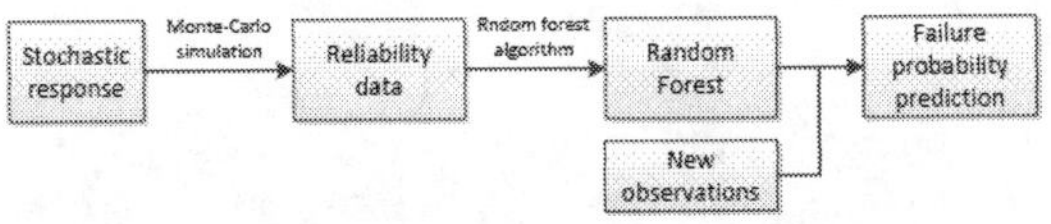

Figure 2 The schematic of the proposed method

3.1 Reliability evaluation by MCS

To apply MCS, the stochastic excitation is specified as a certain number of input random variables [6], X=(X_1, X_2,..., X_n) which can be simulated as a generation of the excitation. N samples X are obtained from a multi-variable Gaussian distribution. The Monte Carlo estimator of the structure failure probability is [7]

$$\hat{P}_f = \frac{1}{N_{mc}} \sum_{i=1}^{N_{mc}} I_f(\mathbf{Z}^{(i)}) \tag{3}$$

where $\mathbf{Z}^{(i)}$ is the ith sample of the excitation process; N_{mc} is the number of samples used in MCS, and $I_f(\cdot)$ $I_f(\cdot)$ satisfies $I_f(\cdot)=1$, if $y_{\max} > y_{\text{limit}}$; $I_f(\cdot)=0$, if $y_{\max} \le y_{\text{limit}}$. Here, $y_{\max} = \max(|\mathbf{y}_s|)$ is the maximum response of the base structure.

2.2 Reliability modeling and prediction by RF

A Random Forest (RF) is a meta-learner comprised of many individual trees called classification and regression tree (CART). A tree has a binary recursive structure. Its learning process is actually the node selecting and splitting process. Each split results in two subsets of the data that falls into this node so that the resulting child nodes are the "purest".

The splitting criterion is used to find the optimal split point. At node t, the best split s is chosen to maximize the goodness of split Δi(s, t).

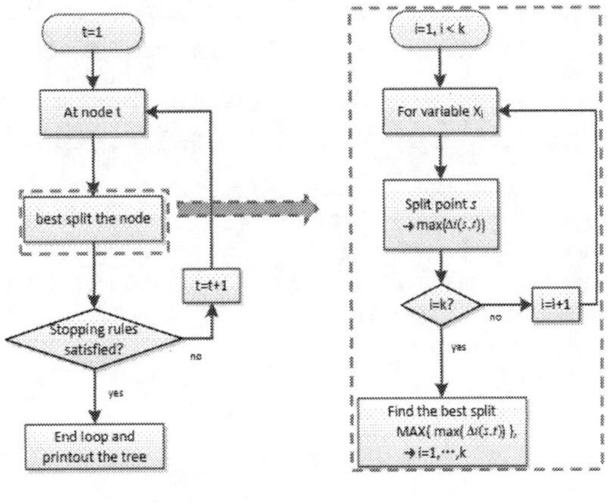

Figure 3 Tree learning process

The goodness of split corresponds to a decrease in impurity. We use Least Squares Deviation (LSD) to measure the impurity. Assume $h(t)$ is the sample set that falls into node t, a is a sample in $h(t)$, then at node t the impurity is defined as

$$i(t) = \sum_{a \in h(t)} (y_a - \bar{y}(t))^2 \tag{4}$$

where y_a is the decision value for sample a, $\bar{y}(t)$ is the average of all decision values in $h(t)$. At node t, the best splitting point s maximizes

$$\Delta i(s,t) = i(t) - i(t_L) - i(t_R) \tag{5}$$

where t_L and t_R are the left child node and right child node produced by the splitting point s. The splitting procedures should be stopped if the defined stopping rules are satisfied.

After training the RF, we determine the prediction value by averaging all predicted values of the individual trees on a new sample $\mathbf{x}'$:

$$\tilde{f} = 1/K \cdot \sum_{b=1}^{K} \tilde{f}_b(\mathbf{x}') \tag{6}$$

where K is the size of RF.

4 Numerical test

In this section, the RF performance in reliability modeling and predictions of an exemplary structure are shown and discussed. The structure parameter configurations come from Elyes [8], see Table 1. Besides, the natural frequency and damping ratio of the seismic model are $\omega_f = 25.224\ rad/s$, $\xi_f = 0.4$. The power spectral density of the white noise process is $S_0 = 0.031W/Hz$. The failure criterion is 4.3e^{-2}m.

Table 1. Parameters of the structure

Parameters	$m(kg)$	$c(N\cdot m/s)$	$k(N/m)$
Base structure	1	0.03	696.4
TMD	0.02	0.0695	12.725
standard dev.	0. 1	0.001	5.0

In this example, the uncertainties exist in the base structure properties. Different sizes of data of the uncertain property values are sampled according to their uncertain characteristics. The sizes of data range from 1000 cases to 19000 cases.

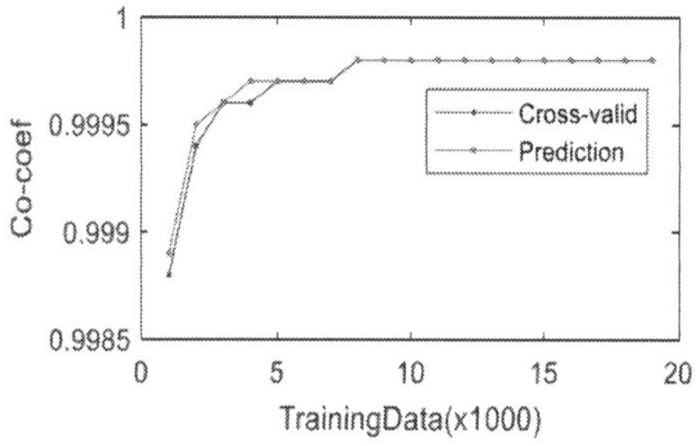

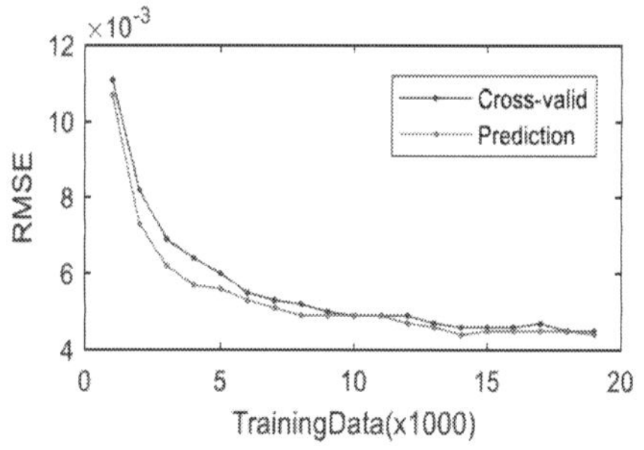

Figure 3 Simulation results based on RF model

We evaluate the model itself by cross-validation on the datasets. Another 1000 cases are sampled as a test set to study the model performance in predictions. From Figure 5, we see that with the increase of the training data, the Co-coef (correlation coefficients) increase no matter in the cross-validation or prediction procedures. This means that the RF model fits the data very well, and more importantly it doesn't overfit the data. Meanwhile, the RMSEs (root mean square error) decrease drastically, which indicates that the model becomes more accurate when more samples are available.

5 Conclusions

In this paper, the Random Forest method is investigated on reliability modeling and prediction of passive controlled structures. Results from the numerical example showed that the RF model behaves well in failure modeling and prediction of passive controlled structures. The RF model is an alternative way to study structure reliability and more attention should be paid to this approach.

References

1. Xu K, Xie M, Tang L C, et al. Appl. Soft Compu., 2003, 2(4): 255-268.
2. Das Chagas Moura M, Zio E, Lins I D, et al. Reliab. Eng. & Sys. Saf., 96, 11(2011).
3. Kutyłowska M. Eng. Failure Ana., 47(2015).
4. Nitze I, Schulthess U, Asche H. Proc. of the 4th GEOBIA (2012).
5. Raczko E, Zagajewski B. European Journal of Remote Sensing, , 50,1(2017).
6. Au S K, Beck J L. Prob. Eng. Mech., 2001, 16(3): 193-207.
7. Yu H, Gillot F, Ichchou M. Jour. of Sound and Vib., 2013, 332(9): 2222-2238.
8. Mrabet E, Guedri M, Ichchou M N, et al. Mech. Sys. and Sig. Proce., 2015, 60: 437-451.

Design of a quasi-periodic vibration energy harvester based on an electromagnetic technique

Zakaria Zergoune*, Najib Kacem and Noureddine Bouhaddi

University Bourgogne Franche-Comté, FEMTO-ST Institute, CNRS/UFC/ENSMM/UTBM, Department of Applied Mechanics, 25000 Besançon, France

Abstract. In the present paper, a quasi-periodic vibration energy harvester with magnetic coupling is proposed using the benefits of the energy localization. The proposed quasi-periodic system consists of moving magnets held by elastic structures and coupled by a magnetic force. The mistuning of the device can be achieved by changing either the linear mechanical stiffness or the mass. The whole system has been modelled by forced Duffing equations for each degree of freedom, which include the magnetic nonlinearity and the mechanical damping. The governing equations have then been solved using the harmonic balance method coupled with the asymptotic numerical method. The obtained numerical results show that the total harvested power was increased by 11 % with a bandwidth of 2.7 % thanks to the energy localization phenomenon compared to the reference case.

1 INTRODUCTION

Embedded systems have gained accuracy, reliability, robustness while becoming miniaturized but they are still limited by the life cycle of their power system. Over the last few years, a great interest for scientific research is showing in a promising alternative to ensure the energy autonomy of nomadic devices by scavenging the ambient mechanical energy. This approach has a strong potential for innovation, miniaturization, respect for ecological issues and is part of the theme of renewable energies as well.

In our environment, diverse ambient energy sources are available and their conversion into electrical energy is a major challenge to increase the autonomy of isolated or abandoned systems. Each environment can correspond to one or more energy sources such as light, wind, thermal gradients, mechanical vibrations. For each of these sources, one or more conversion principles exist for generating electricity. Mechanical vibration sources provide potential energy that can be scavenged for charging self-powered systems. In the last decade, several design of energy harvesters, based on different conversion mechanisms, have been proposed [1,2]. Vibration energy harvesters are commonly based on three different conversion techniques, namely piezoelectric [3,4], electromagnetic [5,6], and for MEMS technology electrostatic [7].

The bulk of the VEH devices proposed in the literature have a chief drawback in that they are effective in a restricted frequency bandwidth around their resonance frequency. Several techniques have been reported in order to enhance the energy harvesting performances. Namely, the multimodal methods and nonlinear have been widely studied [8-10] for enlarging the effective frequency bandwidth and giving more reliable results.

The proposed study deals with the design and modelling of a novel quasi-periodic nonlinear VEH under a harmonic base excitation. The designed harvesting devise used the energy localization and the nonlinear vibration to enhance the harvested performance. Elastic suspensions attached to the moving magnets were used to guarantee a perfect vertical displacement and to reduce the mechanical damping as well. The proposed harvester, operating at low frequencies, demonstrated high performances in term of bandwidth and harvested power by combining the benefits of the low mechanical damping, the energy localization phenomenon and the nonlinearity.

2 SYSTEM MODELING

In the present section, two middle moving magnets are considered as illustrated in the equivalent mechanical and electrical model (Figure 1). The proposed harvesting devise is modelled using two forced duffing equations. So, the governing equation of the designed harvester can be written as:

$$m\ddot{x}_j + c\dot{x}_j + F_j^{me}(x) + F_j^{mg}(x) = -m\ddot{Y} \qquad (1)$$

where $j = 1,2$ and $c = c_e + c_m$ with c_e and c_m stand for the mechanical and electrical damping factors respectively. F_j^{me} and F_j^{mg} are respectively the mechanical and magnetic forces for each moving magnet.

* Corresponding author: zergoune.uni@gmail.com

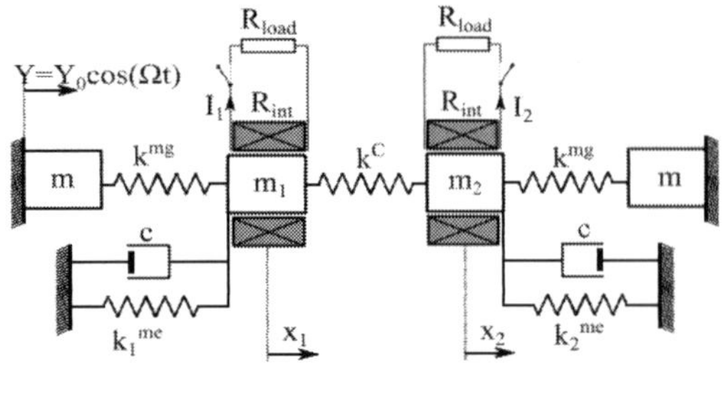

Fig. 1. Equivalent electrical and mechanical model for the proposed VEH device.

It is assumed that the two middle moving magnets have the same mass, size, and the identical mechanical and electrical damping.

The mechanical force, describes the behaviour of the used elastic spring, was modelled by Ansys software in the static model with considering the geometrical nonlinearity.

Figure 2 illustrates the numerical results obtained for the used elastic spring in function of the relative displacement. The obtained results were fitted in order to determine the linear and cubic nonlinear stiffness coefficients k_1^{me} and k_3^{me}. As shown in this figure, the numerical and fitted results are in good agreement. Considering the proposed design, the maximal amplitude cannot exceed 5 mm. In this sense, the linear expression can describe the behaviour of the elastic spring. Therefore, the cubic nonlinear stiffness k_3^{me} is neglected.

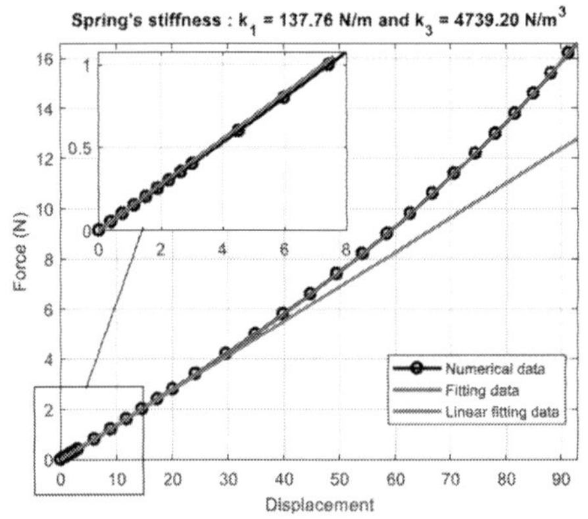

Fig. 2. Mechanical force with respect to the relative displacement.

On the other hand, the repulsive magnetic force has been estimated by the semi-analytical method called the equivalent surface current model (ESCM).

$$F(d_0) = \mu_0 \pi (R M_s)^2 \int_0^{\infty} J_1(\varepsilon R)^2 [2e^{-\varepsilon(d_0+h_{mag})} \\ -e^{-\varepsilon.gap} - e^{-\varepsilon(d_0+2h_{mag})}]\varepsilon^{-1} d\varepsilon, \quad (2)$$

where $J_1(\varepsilon R)$ is the 1st order Bessel function. $\mu_0 = 4\pi 10^{-7}\ H.m^{-1}$ stands for the vacuum magnetic permeability. M_s is the saturation magnetization of the identical magnets. R and h_{mag} are the radius and height, of the magnets.

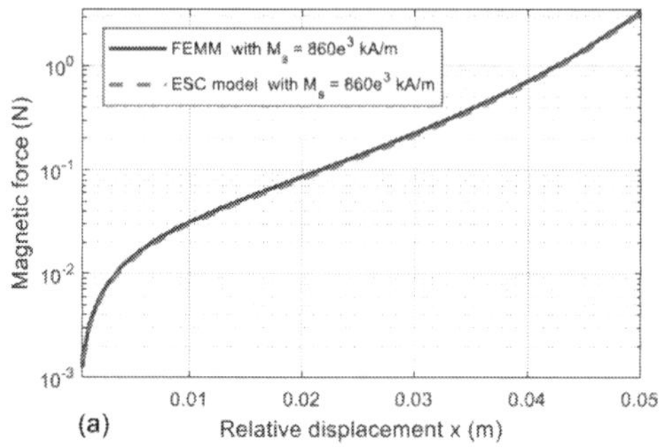

Fig. 3. Comparison of the magnetic force estimated by ESCM and FEMM with $d_0 = 60\ mm$.

The accuracy of the results obtained by the ESC model was checked with the open numerical software FEMM [11]. As shown in Figure 3, the both models are in good agreement. The obtained results of the magnetic force estimated by ESCM have been fitted at different values of the gap d_0 and the relative displacement x using a least-squares procedure. So, the total magnetic force can be identified as:

$$F_j^{mg}(x) = k_1^{mg} x + k_3^{mg} x^3, \quad (3)$$

where $k_1^{mg} = 2\lambda_1 + 4d_0\lambda_2 + 6d_0^2\lambda_3$ is the linear stiffness coefficient and $k_3^{mg} = 2\alpha_3$ is the cubic nonlinear stiffness coefficient in which d_0 is the gap between the magnets. Figure 4 exhibits the fitted data of the total magnetic force with respect to the relative displacement. The fitted data can describe the total magnetic force up to 30% of the gap compared to the ESC model.

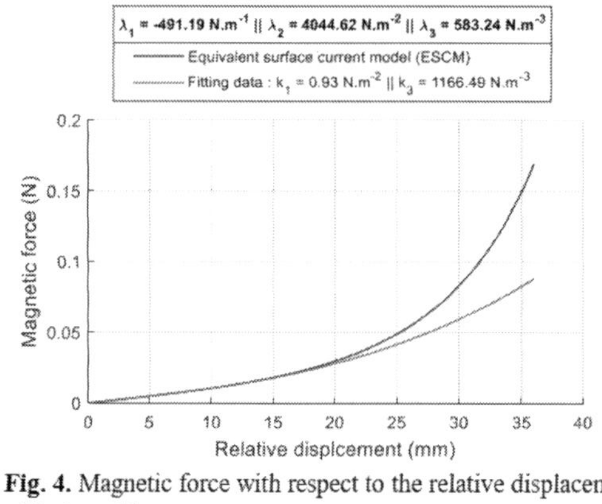

Fig. 4. Magnetic force with respect to the relative displacement with $d_0 = 60\ mm$.

The estimated parameters for the magnetic linear stiffness at $d_0 = 60\ mm$ are $\lambda_1 = -491.19\ N.m^{-1}$, $\lambda_2 = 4044.62\ N.m^{-2}$, and $\lambda_3 = 583.24\ N.m^{-3}$.

The governing equations of the proposed VEH device become as follows by substituting the expression of the magnetic and mechanical forces into equation 1:

$$\begin{cases} \ddot{x}_1 + 2(\xi_e+\xi_m)\omega_1\dot{x}_1 + \omega_1^2(1+2\beta)x_1 - \beta x_2 \\ \quad -\beta_{NL}x_2^3 = -\ddot{Y} \\ \alpha\ddot{x}_2 + 2(\xi_e+\xi_m)\omega_1\dot{x}_2 + \omega_1^2(1+2\beta)x_2 - \beta x_1 \\ \quad -\beta_{NL}x_1^3 = -\alpha\ddot{Y} \end{cases} \tag{4}$$

where $2\xi m_1\omega_1 = c, \beta = \frac{k_c^L}{k_1^{me}}, \beta_{NL} = \frac{k_c^{NL}}{m_1}, \omega_1^2 = \frac{k_1^{me}}{m_1}, \alpha = \frac{m_2}{m_1}$, . α and β are the mass mistuning and coupling coefficients, respectively.

The solving procedure uses the classical harmonic balance method combined with the asymptotic numerical method [12]. This technique allows transforming the nonlinearity present in the governing equation (equation 4) into purely polynomial quadratic terms.

The magnetic field B of the permanent magnets has been obtained analytically by the expression developed for cylindrical magnets [13]:

$$B(x) = \frac{\mu_0 M_s}{2}\left(\frac{x}{\sqrt{x^2+R^2}} - \frac{x-h_{mag}}{\sqrt{(x-h_{mag})^2+R^2}}\right) \tag{5}$$

where x is the relative displacement and R is the radius of the moving magnets. The electromagnetic coupling γ was estimated by the following expression [14]:

$$\gamma = 2N\pi R \int \frac{B(x,h_{mag},R)}{dx}\, dh \tag{6}$$

where $N = 70$ is the number of winding of the coil. The calculated electromagnetic coupling $\gamma = 1.019\ V.s.m^{-1}$ has been compared to the one $\gamma = 1.089\ V.s.m^{-1}$ estimated experimentally by measuring the output voltage at different load resistance values R_{load}.
The average of the total harvested power can be expressed by the following formula:

$$P_{avg} = \left(\frac{\gamma}{\sqrt{2}(R_{load}+R_{int})}\right)^2 R_{load}\,(\dot{x}_1^2 + \dot{x}_2^2) \tag{7}$$

where R_{load} and R_{int} are the load and inner resistances. $\dot{x}_1$ and $\dot{x}_2$ are the velocities of the first and second dofs.

3 RESULTS AND DISCUSSION

In the present section, several numerical simulations have been performed in the case of two moving magnets. These simulations enable us to highlight the benefits of the energy localization phenomenon on the proposed harvesting device.

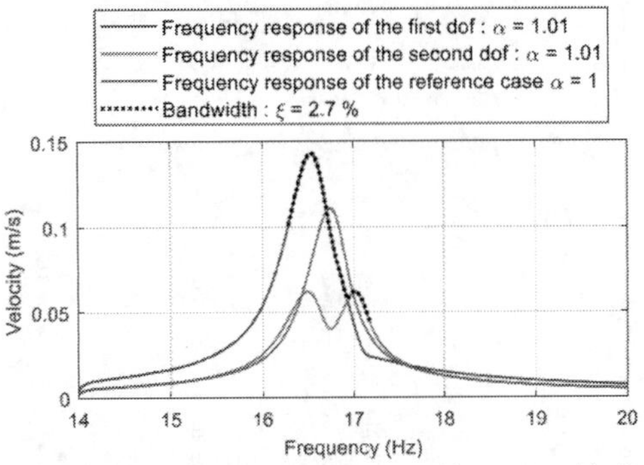

Fig. 5. Frequency response of the proposed harvester with $d_0 = 60\ mm$.

Figure 5 exhibits the frequency response of the proposed VEH device. This later shows an increasing of the velocity of the perturbed dof by approximately 21 %. Also, the effective bandwidth of the VEH device was enlarged by 14 % compared to the reference case.

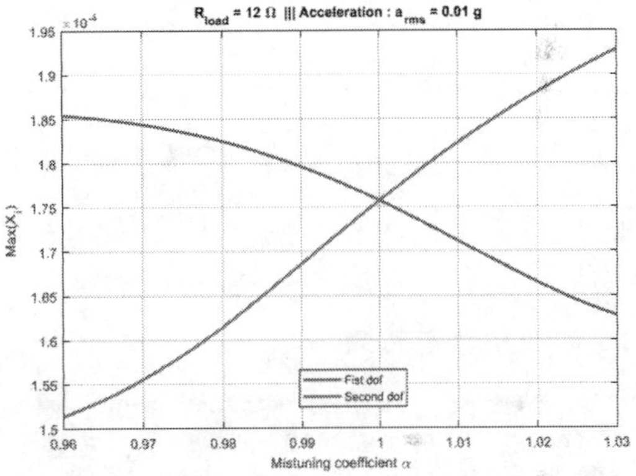

Fig. 6. Maximal oscillation amplitude at different mass mistuning coefficient values α with $d_0 = 60\ mm$.

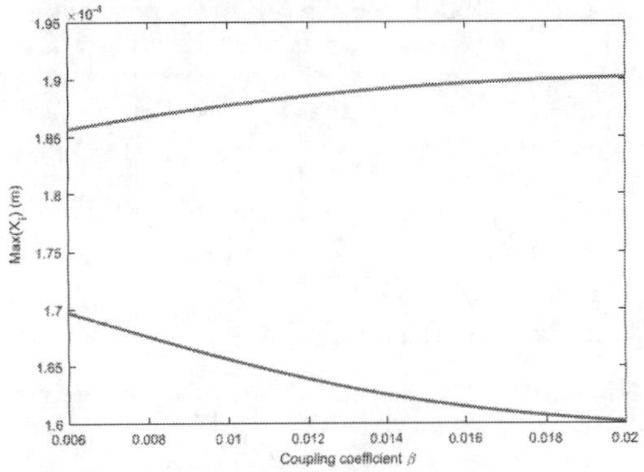

Fig. 7. Maximal oscillation amplitude at different magnetic coupling values β with $d_0 = 60\ mm$.

Figure 6 and 7 represent the variation of the maximal oscillation amplitudes at different values of the mass mistuning and magnetic coupling coefficients α and β. It can be noticed in Figure 6 that an increasing of the oscillation amplitudes of the proposed VEH device compared the case with $\alpha = 1$. In addition, the magnetic coupling coefficient β shows an effect on the oscillation amplitudes as shown in Figure 7.

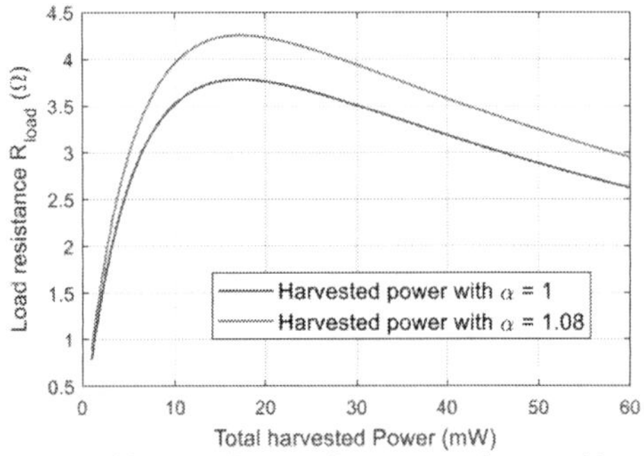

Fig. 7. Total harvested power for $\alpha = 1$ and 1.08 with $d_0 = 60\ mm$.

Figure 7 exhibits the maximal powers at different load resistance with a gap $d_0 = 60\ mm$ harvested from the proposed VEH device in two cases $\alpha = 1$ and 1.08. The estimated magnetic coupling for this gap is $\beta = 0.67$ %. In the first case, the maximal power has been harvested from the both dofs without any perturbation $\alpha = 1$ while in the second case the proposed VEH device was perturbed $\alpha = 1.08$. As shown in this Figure, the harvested maximal power with $\alpha = 1.08$ was increased approximately 11 % compared to the first case with $\alpha = 1$.

4 CONCLUSIONS

In the present work, the effect of the energy localization phenomenon and the magnetic coupling on the frequency response of a periodic structure have been studied. The obtained results show that the perturbation of one of the moving magnet and the magnetic coupling coefficient β increase the oscillation amplitude of the periodic system. Thus, we can take advantage of these aspects to enhance the harvested power of a vibration energy harvesting mechanism. As mentioned before, the total harvested power was improved approximately by 11 % compared to the case without any perturbation. Also, thanks to the energy localization phenomenon, the effective bandwidth was enlarged by 14 %.

ACKNOWLEDGEMENTS

This project has been performed in cooperation with the Labex ACTION program (contract ANR-11-LABX-01-01).

References

1. C. Wei and X. Jing, Renewable and Sustainable Energy Reviews 74, 1 (2017).
2. N. Tran, M. H. Ghayesh, and M. Arjomandi, International Journal of Engineering Science, 127, 162-185 (2018).
3. A. Toprak and O. Tigli, Applied Physics Reviews 1, 031104 (2014).
4. S. Priya, H. C. Song, Y. Zhou, R. Varghese, A. Chopra, S. G. Kim, I. Kanno, L. Wu, D. S. Ha, J. Ryu, and R. G. Polcawich, Energy Harvesting and Systems 4, 3 (2017).
5. Z. Wu, J. Tang, X. Zhang, and Z. Yu, Applied Physics Letters 111, 013903 (2017).
6. S. Dirk and M. Yiannos, Electromagnetic Vibration Energy Harvesting Devices (Springer, 2012).
7. Y. Zhang, T. Wang, A. Zhang, Z. Peng, D. Luo, R. Chen, and F. Wang, Review of Scientific Instruments 87, 125001 (2016).
8. S. Mahmoudi, N. Kacem and N. Bouhaddi, Enhancement of the performance of a hybrid nonlinear vibration energy harvester based on piezoelectric and electromagnetic transductions, Smart Materials and Structures 23, 075024 (2014).
9. I. Abed, N. Kacem, N. Bouhaddi and M. L. Bouazizi, Multi-modal vibration energy harvesting approach based on nonlinear oscillator arrays under magnetic levitation. Smart Materials and Structures, 25, 025018 (2016).
10. C. Drezet, N. Kacem and N. Bouhaddi, Design of a nonlinear energy harvester based on high static low dynamic stiffness for low frequency random vibrations, Sensors and Actuators A: Physical (2018).
11. D. C. Meeker, http://www.femm.inf (2006).
12. B. Cochelin and C. Vergez, Journal of Sound and Vibration, 324, 243-262 (2009).
13. J. M. Camacho, V. Sosa, Revista Mexicana de Fisica, 59, 8-17 (2013).
14. S. Dirk and M. Yiannos, Electromagnetic Vibration Energy Harvesting Devices, Springer (2012).

Bifurcation, chaotic and hysteresis phenomena of broadband tristable energy harvesters

Shengxi Zhou[1], Junyi Cao[2], and Grzegorz Litak[3,4,a]

[1]School of Aeronautics, Northwestern Polytechnical University, Xi'an, 710072, China
[2]School of Mechanical Engineering, Xi'an Jiaotong University, Xi'an, 710049, China
[3]Lublin University of Technology, Faculty of Mechanical Engineering, Nadbystrzycka 36, PL-20-618 Lublin, Poland
[4]Department of Process Control, AGH University Science and Technology, Mickiewicza 30, PL-30-059 Krakow, Poland

Abstract. We analyze energy harvesting using a mechanical resonator with three potential wells. Nonlinear effects are leading to frequency broadband voltage response via bifurcations, multiple solutions and dynamical hysteresis. We show the characteristic features of the obtained solutions. In particular, basins of attraction of solutions are discussed.

1 Introduction

Vibration energy harvesting has been expected to be a promising way to solve the challenging issue of the permanent energy supply for small sensors and MEMS devices [1–5]. By far, the response characteristics and energy harvesting performance of nonlinear monostable and bistable energy harvesters have been widely revealed and verified [6–13]. Among currently existing energy harvesters, the tristable energy harvester (TEH) is one of most recent energy harvesters [14–17]. More importantly, the TEH was theoretically and experimentally verified that it has very good energy harvesting performance under low-frequency and low-level excitations. However, there is still missing deep investigation on the dynamic response of the TEH. Therefore, this paper explores the bifurcation, chaotic and hysteresis phenomena of broadband TEHs based on simulation and experiments. In detail, the bifurcation phenomena of the TEH are analyzed by using Poincaré maps and the phase portrait topology. The responses of the TEH under linearly increasing frequency sweep excitation and decreasing frequency sweep excitation clearly demonstrate its existing dynamical hysteresis phenomena (Fig. 1).

The schematic diagram of the TEH is shown in Fig. 1. The harvester mainly includes a piezoelectric beam, several magnets, a load resistance and a supporting mechanism. The nonlinear magnetic force can be produced by the interaction among endmost magnets and external magnets, which depends on the relative position of these magnets. Therefore, the equivalent nonlinear restoring force is the vector sum of the linear restoring force of the piezoelectric beam and the nonlinear magnetic force. The TEH has five equilibrium positions (1, 3 and 5 are stable equilibrium positions, while 2 and 4 are unstable equilibrium positions) [14-17]. Under the base displacement excitation, the governing model of the THE can be described by the following equations [16]:

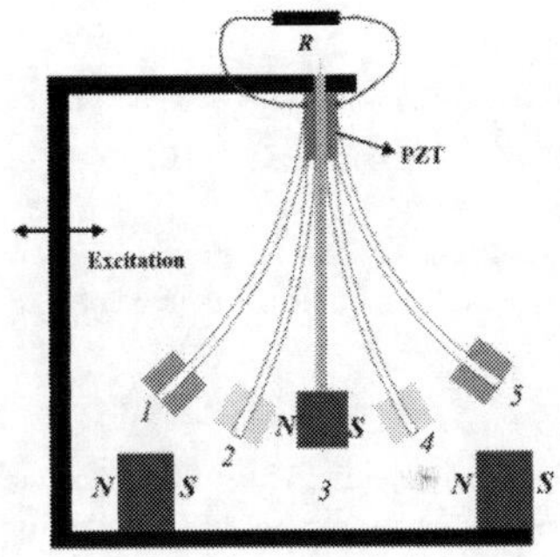

Figure 1. Schematic diagram of the tristable energy harvester.

$$m\ddot{y}(t) + c\dot{y}(t) - F_r(y) - \theta v(t) = F(t), \qquad (1)$$

$$C_p\dot{v}(t) + v(t)/R + \theta\dot{y}(t) = 0, \qquad (2)$$

where m and c are the equivalent mass and the equivalent damping, respectively; $y(t)$ is the tip horizontal displacement relative to the base dependent on time; C_p, R and θ are the equivalent capacitance, the load resistance and the equivalent electromechanical coupling coefficient, respectively; $V(t)$ is the output voltage (it is also the voltage at the load resistance R); F_r in the equivalent nonlinear restoring force, with accuracy to the sign, while $F(t)$ is the modal excitation force.

This paper mainly focuses on investigating the general dynamic response characteristics of the TEH for enhancing vibration energy harvesting performance. Therefore, a dimensionless model of the TEH is studied. Under a

[a]e-mail: g.litak@pollub.pl

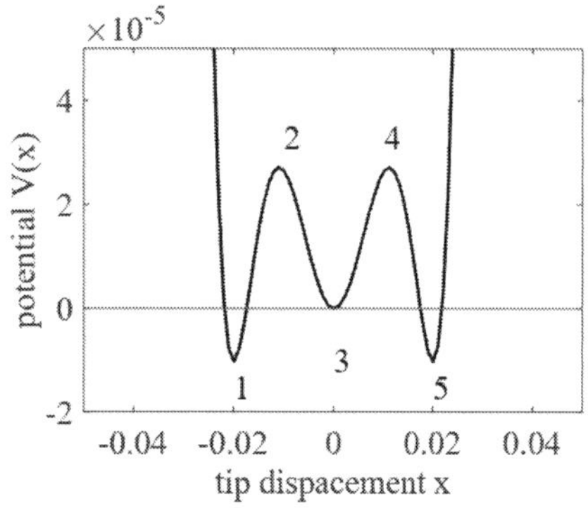

Figure 2. Potential of the examined TEH system with a tristable characteristic. Numbers 1-5 correspond to stable (1,3,5) and unstable (2,4) equilibria, respectively.

harmonic base acceleration excitation $A\cos(\omega t)$, the dimensionless dynamic model of the TEH can be described by the following equations [16]:

$$\ddot{x}(t) + \xi\dot{x}(t) - F_r(x) - \theta v(t) = A\cos(\omega t), \quad (3)$$

$$C_p\dot{v}(t) + v(t)/R + \theta\dot{x}(t) = 0, \quad (4)$$

where ξ is the dimensionless damping coefficient. $F_r(x)$, dependent on the dimensionless displacement x can be expressed as:

$$F_r = -(x - \mu x^3 + \beta x^5) \quad (5)$$

where μ and β are the coefficients of the cubic x^3 and quintic x^5 terms of the equivalent nonlinear restoring force, respectively; the coefficient of x is set to be one for analysis. In order to make five roots (to yield a tristable characteristic), F_r should meet a specific condition which can be deduced based on Eq. (5). Then, the potential energy function can be expressed as (see also in Fig. 2):

$$V(x) = \frac{1}{2}x^2 - \frac{1}{4}\mu x^4 + \frac{1}{6}\beta x^6. \quad (6)$$

It is noted that all the parameters are dimensionless for numerical analysis, while these parameters can be theoretically calculated or experimentally measured based on a specific energy harvesting device. In this study, the dimensionless system parameters are set to: ξ =0.8, C_p =0.05, θ =0.5, R = 1.0×10^7 , μ = −1.0764×10^4, β = 2.0661×10^7.

2 Basins of attraction of the competing solutions

Basins of attraction of TEH are explored below. In this paper, coexisting orbits are named as interwell attractor and intrawell attractor, and their basins of attraction are depicted in white and black colors, respectively. Chaotic, quasi-periodic, or harmonic motions were not specifically distinguished. Basins of attraction were obtained based on the numerical calculation from these initial conditions via

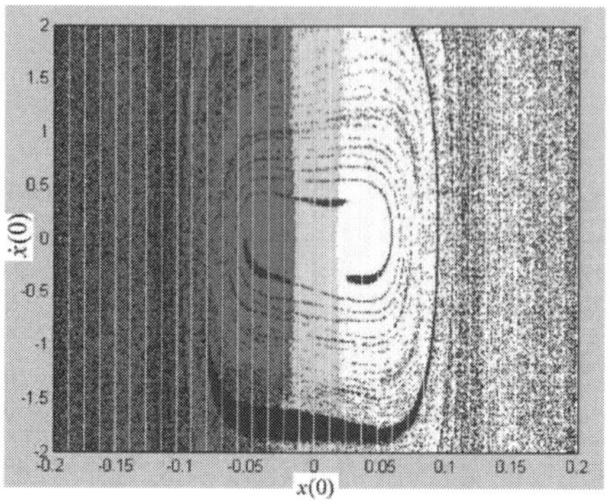

Figure 3. Basins of attraction: the excitation frequency f = 8π and the excitation amplitude A = 1.2. Red, green and yellow colours correspond to basins of intrawell oscillatory solutions around equlibria 1,3,5, respectively, while blue correspond to basin of interwell oscillations.

MATLAB software. Initial velocity and initial displacement were varied from -2.0 to 2.0 and -0.2 to 0.2, respectively. The interwell oscillation phase, and the intrawell oscillation phases around equilibrium point 1, 3 and 5 are depicted in blue, red, green and yellow colors, respectively. In Fig. 3, the excitation frequency ω = 8π and the excitation amplitude A = 1.2 were selected. It was found that 20.8% of initial conditions yield high-energy interwell oscillations. The proportion of initial conditions yielding intrawell oscillations equilibrium point 1, 3 and 5 is 33.8%, 11.2% and 34.2%, respectively. Keeping A = 1.2 and increasing ω to 9π the proportion of initial conditions which yield high-energy interwell oscillations decreased to 4.8%, as shown in Fig. 3. Meanwhile, the proportion of initial conditions yielding intrawell oscillations equilibrium point 1, 3 and 5 is increased to 41.1%, 13.1% and 41.0% respectively. When ω is further increase to 10π (with the same A = 1.2), only 2.1% initial conditions yield high-energy interwell oscillations, as shown in Fig. 4. The proportion of initial conditions yielding intrawell oscillations equilibrium point 1 and 5 becomes exactly the same and is 42.9%. This tendency indicates that the TEH can be much easier to achieve high-energy interwell oscillations under a low excitation frequency with the same excitation amplitude. It is also worth to notice that the solutions are depicted with different colours (red, green, yellow and blue). The basins demonstrate fractal borders indicating existence of the chaotic solutions. Fractal mixing of three basins of attraction indicate the Wada patterns [18, 19].

3 Experimental results: dynamical hysteresis

For the assumed system (Fig. 1) we performed experimental studies. As analyzed above, the TEH has numerical multi-solution ranges and hysteresis phenomenon

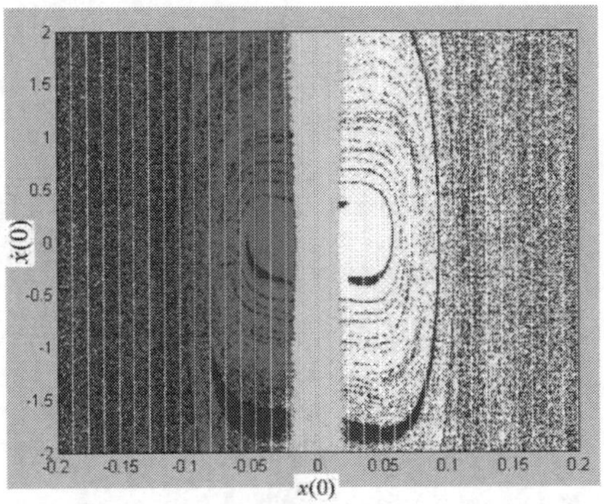

Figure 4. Basins of attraction: the excitation frequency $f = 9\pi$ and the excitation amplitude $A = 1.2$. Red, green and yellow colours correspond to basins of intrawell oscillatory solutions around equlibria 1,3,5, respectively, while blue correspond to basin of interwell oscillations.

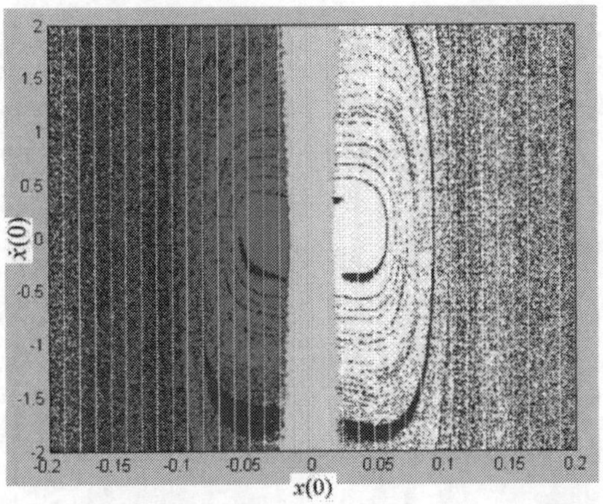

Figure 5. Basins of attraction: the excitation frequency $f = 10\pi$ and the excitation amplitude $A = 1.2$. Red, green and yellow colours correspond to basins of intrawell oscillatory solutions around equlibria 1,3,5, respectively, while blue correspond to basin of interwell oscillations.

corresponding to the excitation frequency and amplitude. To confirm these characteristics, linearly increasing frequency-swept excitation (up-sweep) and linearly decreasing frequency-swept excitation (down-sweep) experiments under the excitation level of acceleration fixed 0.27 g and 0.35 g (where g is the Earth gravitation acceleration) depicted in Fig. 6a and b, respectively. At 0.27 g, the TEH has hysteresis phenomenon (interwell and intrawell orbits) in the frequency range of 7.8-15.9 Hz. At 0.35 g, the corresponding frequency range shifts to 9-17.1 Hz. Therefore, the experiments qualitatively verify the numerical conclusion.

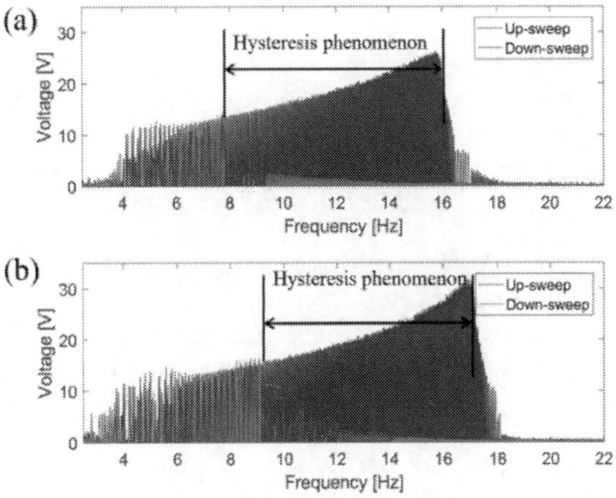

Figure 6. Dynamical hysteresis phenomenon for voltage output of the TEH system in experimental results under linearly increasing frequency sweep excitation (Up-sweep) and decreasing frequency sweep excitation (Down-sweep). (a) and (b) corresponds to different levels of harmonic excitation.

4 Conclusions

The main feature of the studied system is the appearance of additional solutions absent in the linear system. This gives the possibilities to adjust various solutions closer to resonance area. Consequently, by changing frequency we observe the specific fitting of the rational frequency response to the frequency of excitation leading to the larger interwell oscillation amplitude of the voltage output. Note that, a chaotic oscillation is also blonging to such a class of interwell solutions. The dynamical hysteresis shown in Fig. 6 is related to the switching between two or more solutions (see Figs. 3-5) during the excitation with variable frequencies. The occurrence of the multiple solutions at a given frequency and also dynamical hysteresis during the frequency sweep give the opportunity to broaden effectively the excitation frequency width of the TEH. This is the most important characteristics in the ambient vibrational energy source where the amplitude and frequency fluctuate.

References

[1] P.D. Mitcheson, E.M. Yeatman, G.K. Rao, A.S. Holmes, T.C. Green, Proceedings of the IEEE **96**, 1457 (2008)
[2] R.L. Harne, K.W. Wang, Smart Mat. Struct. **22**, 023001 (2013)
[3] S.P. Pellegrini, N. Tolou, M. Schenk, J.L. Herder, J. Intell. Mater. Syst. Struct. **24**, 1303 (2013)
[4] J. Twiefel, H. Westermann, J. Intell. Mater. Syst. Struct. **24**, 1291 (2013)
[5] S.P. Beeby, M.J. Tudor, N.M. White, Meas. Sci. Technol. **17**, R175 (2006)
[6] S. Zhou, J. Cao, A. Erturk, J. Lin, Appl. Phys. Lett. **102**, 173901 (2013)

[7] A. Syta, G. Litak, M.I. Friswell, S. Adhikari, Eur. Phys. J. B **89**, 99 (2016)

[8] T. Huguet, A. Badel, A., M. Lallart, Appl. Phys. Lett. 111, 173905 (2017)

[9] M.I. Friswell, S.F. Ali, S. Adhikari, A.W. Lees, O. Bilgen, G. Litak, J. Intell. Mater. Syst. Struct. **23**, 1505 (2012)

[10] A. Erturk, J. Hoffmann, D.J. Inman, Appl. Phys. Lett. **94**, 254102 (2009)

[11] F. Cottone, H. Vocca, L. Gammaitoni, Phys. Rev. Lett. **102**, 080601 (2009)

[12] S. Stoykov, E. Manoach, G. Litak, Eur. Phys. J. Special Topics **224**, 2755 (2015)

[13] M. Borowiec, G. Litak, S. Lenci, Inter. J. Struct. Stab. Dynamics **14**, 1440020 (2014)

[14] S. Zhou, J. Cao, D.J. Inman, J. Lin, S. Liu, Z. Wang. Appl. Energy **133**, 33-39 (2014)

[15] S. Zhou, J. Cao, D.J. Inman, J. Lin, D. Li, J. Sound. Vib **373**, 223 (2016)

[16] S. Zhou, L. Zuo, Commun. Nonlinear. Sci. Numer. Simulat. **61**, 271 (2018)

[17] S. Zhou, J. Cao, G. Litak, J. Lin, TM-Tech. Mess. **85**, 521-532 (2018)

[18] J. Aguirre, J.C. Vallejo, M.A.F. Sanjuan, Phys. Rev. E **64**, 066208 (2001)

[19] G. Litak, M. Coccolo, M.I. Friswell, S.F. Ali, S. Adhikari, A.W. Lees, O. Bilgen, in "Nonlinear oscillations of an elastic inverted pendulum," 2012 IEEE 4th International Conference on Nonlinear Science and Complexity (NSC), Budapest, 2012, pp. 113-116 (2012)

Energy harvesting in a delayed Rayleigh harvester device

I. Kirrou[1,a], A. Bichri[2], M. Belhaq[3]

[1] ENSA-Agadir, University Ibn Zohr, Morocco
[2] FST-Errachidia, University Moulay Ismaïl, Morocco
[3] Laboratory of Renewable Energy and Structural Dynamics, University Hassan II-Casablanca, Morocco

Abstract. In the present work, we report on energy harvesting in a delayed Rayleigh oscillator coupled to a piezoelectric harvester device. Analytical investigation using the multiple scales method is performed to obtain approximation of the periodic and QP amplitude response. The influence of different parameters of the harvesting system on the output power coupling to periodic and QP vibrations is examined. Results show that for appropriate values of the delay parameters, QP vibrations can also be exploited to harvest energy with good performence.

1 Introduction

The efficiency of the energy harvesting (EH) has received great attention in the last few years. The main goal that EH wants to achieve is to reduce the requirements of an external source by considering simplified models of harvester device. One recent idea that has received growing attention is energy harvesting systems operating in a self-induced oscillations regime. In recent paper [1], QP vibration-based EH has been studied in a delayed van der Pol oscillator coupled to an electromagnetic EH device and later by considering a forced and delayed Duffing harvester device [2]. It was concluded that energy can be extracted from QP vibrations over broadband of excitation frequency away from the resonance.
A common method to design a wide band of EH is the inclusion of delay feedback in the position, in the velocity or in both. For instance, it was shown that for appropriate values of time delay and feedback gains the output power can be obtained from QP vibrations with a good performance [2,3]. Note that, the influence of time delayed feedback absorber has been used to enhance EH [4].
The purpose of this work is to study the EH performance in a delayed Rayleigh harvester device subjected to a harmonic excitation. The energy harvester system consists in a delayed Rayleigh oscillator coupled to an electrical circuit through a piezoelectric coupling mechanism. Special attention is paid to the effect of delay and coupling parameters on the EH performance taking advantage of large-amplitude QP vibrations produced in the system.
The paper is organized as follows: in section 2 we investigate periodic and the average power near the primary resonance. Section 3 is devoted to QP vibration-based EH and section 4 concludes the work.

2 Equation of motion and periodic energy harvesting

We consider a harvester device consisting of a nonlinear single spring-masse-damper system subjected to a harmonic excitation near the primary resonance. The schematic model of the system is shown in Fig.1 and the governing equation for the harvester can be written in the dimensionless form as

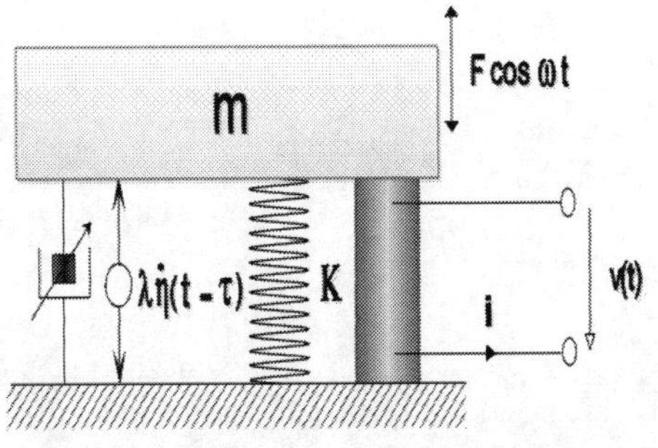

Fig. 1. Description of the energy harvesting system.

$$\ddot{\eta}(t) + \omega_0^2\eta(t) - [\alpha_1\dot{\eta}(t) - \alpha_2\dot{\eta}(t)^2]\dot{\eta}(t) - \theta v(t) = \lambda\dot{\eta}(t-\tau) + Fcos(\omega t)$$
$$\dot{v}(t) + \beta v(t) + \alpha\dot{\eta}(t) = 0 \qquad (1)$$

where $\eta(t)$ is the relative displacement of the rigid mass m, $v(t)$ is the voltage across the load resistance, α_1 and α_2 are, respectively, the linear and nonlinear mechanical damping ratio, θ is the piezoelectric coupling term in the mechanical attachment, α is the piezoelectric coupling term in the electrical circuit, β is the reciprocal of the time constant of electrical circuit, F and ω are, respectively, the amplitude and the frequency of the excitation, while λ and τ are, respectively, the feedback gain and time delay.

To investigate the response of the harvester system (1) near the primary resonance we consider the resonance condition $\omega = \omega_0 + \sigma$ where σ is detuning parameter. The method of multiple scales [5] is implemented by introducing a bookkeeping parameter ε and scaling parameters as $\alpha_1 = \varepsilon\tilde{\alpha}_1$, $\alpha_2 = \varepsilon\tilde{\alpha}_2$, $\theta = \varepsilon\tilde{\theta}$, $\lambda = \varepsilon\tilde{\lambda}$, $F = \varepsilon\tilde{F}$, $\beta = \varepsilon\tilde{\beta}$, $\alpha = \varepsilon\tilde{\alpha}$, $\sigma = \varepsilon\tilde{\sigma}$. Thus, system (1) reads

$$\ddot{\eta}(t) + \omega_0^2\eta(t) = \varepsilon\Big[- \alpha_1\dot{\eta}(t) - \alpha_2\dot{\eta}(t)^3 + \theta v(t) +$$

[a] e-mail: ilhamkirrou@gmail.com

$$\lambda\dot{\eta}(t-\tau) + Fcos(\omega t)\Big]$$
$$\dot{v}(t) + \beta v(t) + \alpha\dot{\eta}(t) = 0 \quad (2)$$

To ultimately solve Eq. (2), steady-state solutions are expanded as

$$\eta(t) = \eta_0(T_0, T_1, T_2) + \varepsilon\eta_1(T_0, T_1, T_2) + \varepsilon^2\eta_2(T_0, T_1, T_2) + O(\varepsilon^3)$$
$$v(t) = v_0(T_0, T_1, T_2) + \varepsilon v_1(T_0, T_1, T_2) + \varepsilon^2 v_2(T_0, T_1, T_2) + O(\varepsilon^3) \quad (3)$$

where $T_0 = \tau$, $T_1 = \epsilon\tau$ and $T_2 = \epsilon^2\tau$. In terms of the variables $T_i(i = 0, 1, 2)$, the time derivatives become $\frac{d}{d\tau} = D_0 + \varepsilon D_1 + \varepsilon^2 D_2 + O(\varepsilon^3)$ and $\frac{d^2}{d\tau^2} = D_0^2 + 2\varepsilon D_0 D_1 + \varepsilon^2 D_1^2 + 2\varepsilon^2 D_0 D_2 + O(\varepsilon^3)$, where $D_i = \frac{\partial}{\partial T_i}$. Substituting (3) into (2), equating terms of different order of ε, resolving and eliminating the secular terms, we obtain the following slow flow of the amplitude and the phase:

$$\begin{cases} \dfrac{dr}{dt} = S_1 r + S_3 r^3 - h\sin\phi \\ r\dfrac{d\phi}{dt} = S_2 r - h\cos\phi \end{cases} \quad (4)$$

where $S_1 = -\frac{\alpha_1}{2} - \frac{D_c}{2\omega} + \frac{\lambda_s}{2\omega}$, $S_2 = \frac{\sigma}{2\omega} - \frac{\lambda_c}{2\omega} - \frac{D_s}{2\omega}$, $S_3 = -\frac{3}{8}\alpha_2\omega^2$, $h = \frac{F}{2\omega}$, $\lambda_c = \frac{\lambda\alpha\omega cos\varphi}{\sqrt{\beta^2+\omega^2}}$, $\lambda_s = \frac{\lambda\alpha\omega sin\varphi}{\sqrt{\beta^2+\omega^2}}$, $D_c = -\lambda\omega cos\omega\tau$ and $D_s = \lambda\omega sin\omega\tau$. Equilibria of this slow flow, corresponding to periodic solutions of Eq. (2), are determined by setting $\frac{dr}{dt} = \frac{d\phi}{dt} = 0$. This leads to the amplitude-frequency response equation:

$$S_3^2 r^6 + 2S_1 S_3 r^4 + (S_1^2 + S_2^2)r^2 - h^2 = 0 \quad (5)$$

An expression for the average power is obtained by integrating the dimensionless form of the instantaneous power $P(t) = \beta v(t)^2$ over the period of the excitation T. This is given by

$$P_{av}^P = \frac{1}{T}\int_0^T \beta v^2 dt \quad (6)$$

where $T = \frac{2\pi}{\lambda}$. Then, the average power expressed by $P_{av}^P = \frac{\beta v^2}{2}$ reads

$$P_{gen}^{av} = \frac{1}{2}\frac{\beta\alpha^2\omega^2}{\beta^2 + \omega^2}r^2 \quad (7)$$

where the amplitude r is obtained from Eq. (5).

3 Quasi-periodic energy harvesting

Next, we use the second perturbation procedures to approximate periodic solutions of the slow flow (4) corresponding to QP responses of Eq. (2) and we analyze the influence of parameters on the QP output power. This can be done by transforming the modulation equations from the polar form (4) to the following Cartesian system using the variable change $u = a\cos\phi$ and $v = -a\sin\phi$

$$\begin{cases} \dfrac{du}{dt} = S_2 v + \mu\{S_1 u + S_3(u^2 + v^2)u\} \\ \dfrac{dv}{dt} = -S_2 u + h + \mu\{S_1 v + S_3(u^2 + v^2)v\} \end{cases} \quad (8)$$

where μ is a new bookkeeping parameter introduced in damping and nonlinearity. Following [1,6], a periodic solution of the slow flow (8) can be sought in the form

$$u(t) = u_0(T_1, T_2) + \mu u_1(T_1, T_2) + O(\mu^2)$$
$$v(t) = v_0(T_1, T_2) + \mu v_1(T_1, T_2) + O(\mu^2) \quad (9)$$

where $T_1 = t$ and $T_2 = \mu t$. Introducing $D_i = \frac{\partial}{\partial T_i}$ yields $\frac{d}{dt} = D_1 + \mu D_2 + O(\mu^2)$, substituting (9) into (8), collecting terms, resolving and eliminating secular terms, we obtain the following autonomous *slow-slow* flow system on R and ψ

$$\frac{dR}{dt} = S_3 R^3 + \left(S_1 + \frac{S_3 h^2}{S_2^2}\right)R$$
$$R\frac{d\psi}{dt} = 0 \quad (10)$$

where R and ψ are, respectively, the amplitude and the phase of the slow flow limit cycle. Equilibria of this *slow-slow* flow corresponding to periodic solutions of the slow flow (8) determine the QP solution of the original equation (2). In addition to the trivial equilibrium, $R = 0$, the non-trivial equilibrium is obtained by setting $\frac{dR}{dt} = 0$ and given by

$$R = -\sqrt{\frac{S_1}{S_3} + \frac{h^2}{S_2^2}} \quad (11)$$

The approximate periodic solution of the slow flow (8) is given by

$$u(t) = R\cos(\kappa t) + \frac{h}{S_2}$$
$$v(t) = -R\sin(\kappa t) \quad (12)$$

The approximate amplitude $a(t)$ of the QP oscillations is given by

$$a(t) = \sqrt{R^2 + h_1^2 + 2h_1 R\cos(\kappa t)} \quad (13)$$

while the modulation envelope of these QP oscillations is delimited by a_{min} and a_{max} given by

$$a_{min} = min\left\{\sqrt{R^2 + h_1^2 \pm 2h_1 R}\right\} \quad (14)$$

$$a_{max} = max\left\{\sqrt{R^2 + h_1^2 \pm 2h_1 R}\right\} \quad (15)$$

The average QP power is obtained from

$$P_{av}^{QP} = \frac{1}{2}\frac{\beta\alpha^2 v^2}{\beta^2 + v^2}a^2 \quad (16)$$

where a is derived from Eqs. (14), (15).
The influence of different system parameters on the maximum output powers is examined. In what follows, we fix the parameters $F = 0.1$, $\omega_0 = 1$, $\theta = 0.3$, $\beta = 0.1$, $\alpha = 0.2$, $\alpha_1 = 0.01$ and $\alpha_2 = 0.01$.
Figures 2, 3 and 4 show, respectively, the variation the output average power amplitudes versus the external frequency ω for fixed values of parameters and for $\lambda = 0$

(Fig.2), λ = 0.07 (Fig.3) and λ = −0.07 (Fig.4). The periodic response is given by (5) and the QP modulation envelope is obtained from Eqs. (18) and (19). Also, the average power for the periodic response is given by Eq. (7), while that for the QP response is deduced from (20). The solid lines represent stable solution and the dashed lines correspond to unstable ones. Figure 3 and 4 indicates

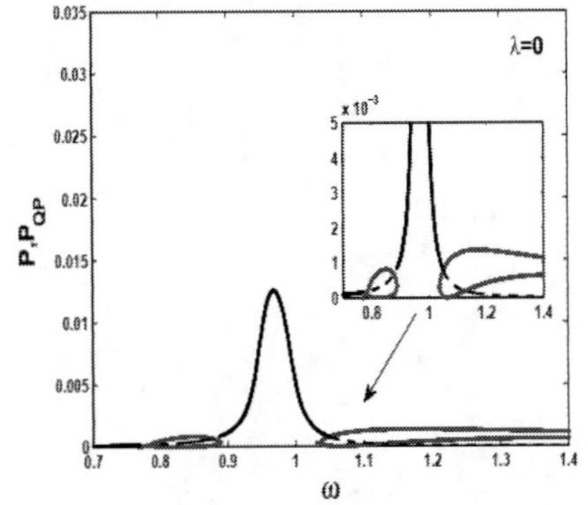

Fig. 2. Vibration and power amplitudes versus ω for $\lambda = 0$; analytical approximation: solid lines for stable and dashed line for unstable solution.

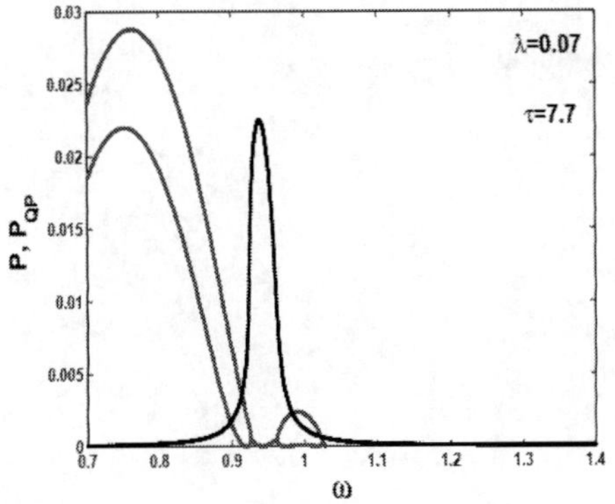

Fig. 3. Vibration and power amplitudes versus ω for $\lambda = 0.07$ and $\tau = 7.7$; analytical approximation: solid lines for stable and dashed line for unstable solution.

that for given values of the delay amplitude λ, the periodic vibration-based EH can be extracted in a small range near the resonance, thereby limiting considerely the bandwidth of the harvester. Instead, QP vibration-based EH can be extracted in relatively broadband of excitation frequencies far from the resonance. Also it can be observed that for positive values of λ, energy extracted from QP domain can be achieved for excitation freuquencies prior to the resonance (left to the resonance peak; Fig 3b), while for negative values of λ QP vibration-based EH can be achieved beyond the resonance (right to the resonance peak; Fig.4).
Figure 5 represents analytical approximations of periodic and QP vibrations and inset in the figure is shown time

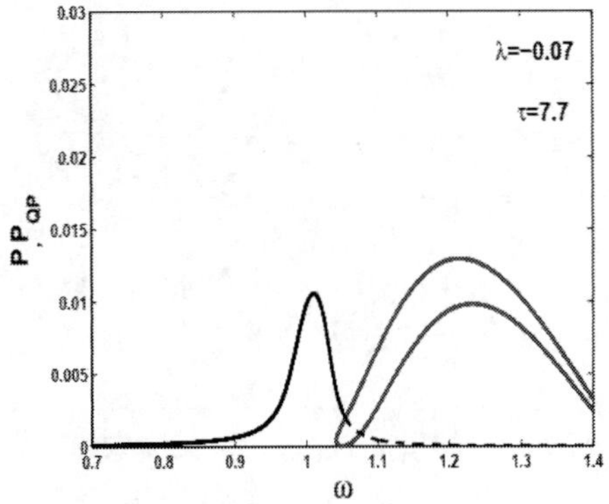

Fig. 4. Vibration and power amplitudes versus ω for $\lambda = -0.07$ and $\tau = 7.7$; analytical approximation: solid lines for stable and dashed line for unstable solution.

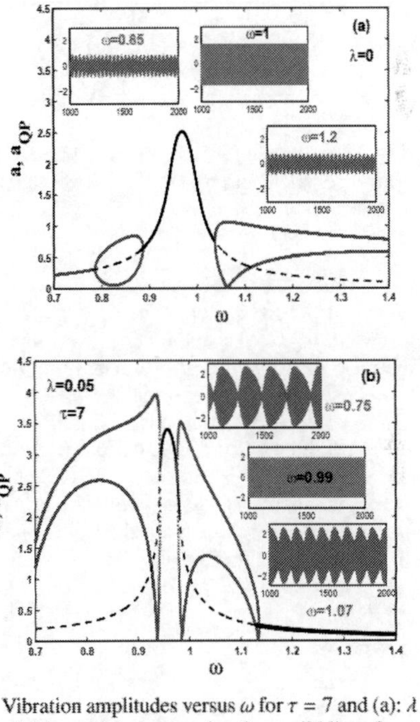

Fig. 5. Vibration amplitudes versus ω for $\tau = 7$ and (a): $\lambda = 0$ and (b): $\lambda = 0.05$; analytical approximation: solid lines for stable and dashed line for unstable solution; time histories corresponding to different regions.

history response using numerical simulations. Figure 6a shows the stability chart of the QP solution in the parameter plane (λ, τ) indicating the green regions where stable QP solutions exist and the gray region corresponding to unstable QP oscillations. In Fig. 6b are shown time histories response related to crosses (a), (b) and (c) picked from Fig. 6a.

4 Conclusions

We have studied EH performance in a delayed Rayleigh oscillator coupled to a piezoelectric harvesting device. The

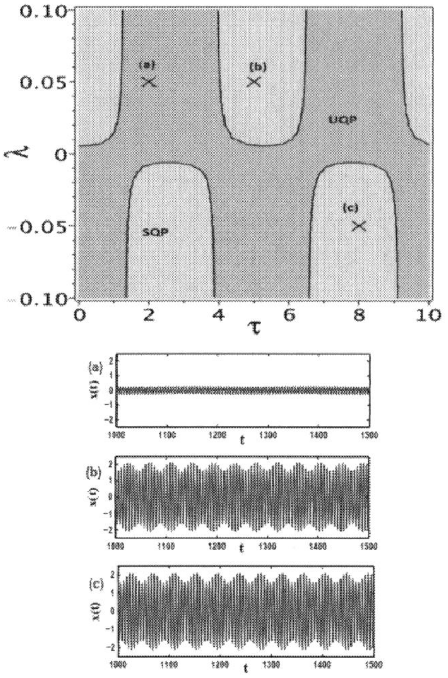

Fig. 6. Stability chart in the plane (λ, τ) and time histories corresponding to different regions; SQP (stable QP solutions); UQP (unstable QP solutions).

method of multiple scales is performed to obtain the slow-slow flows of the EH system. The periodic and QP solutions as well as the corresponding power amplitudes are obtained and the influence of different system parameters on the EH performance is reported. Results shown that for a small value of delay amplitude, the periodic vibration-based EH can be extracted in a narrow range around the resonance, while QP vibration-based EH can be extracted over wide ranges of excitation frequency away from the resonance at the left for positive values of the delay amplitude and at the right for negative ones.

References

[1] M. Belhaq, M. Hamdi, Energy harvesting from quasiperiodic vibrations. Nonlinear Dyn. **86**, 2193-2205 (2016).
[2] Z. Ghouli, M. Hamdi, F. Lakrad, M. Belhaq, Quasiperiodic energy harvesting in a forced and delayed Duffing harvester device. J. Sound Vib. **407**, 271-285 (2017).
[3] Z. Ghouli, M. Hamdi, M. Belhaq, Energy harvesting from quasi-periodic vibrations using electromagnetic coupling with delay. Nonlinear Dyn. **89**, 1625-1636 (2017).
[4] A.S. Kammer, N. Olgac, Delayed feedback vibration absorbers to enhance energy harvesting. Journal of Sound and Vibration **363**, 54-67 (2016).
[5] A.H. Nayfeh, D.T. Mook, Nonlinear Oscillations. Wiley, New York (1979).
[6] M. Belhaq, M. Houssni, Quasi-periodic oscillations, chaos and suppression of chaos in a nonlinear oscillator driven by parametric and external excitations. Nonlinear Dyn. **18**, 1-24 (1999).

Discrete hysteretic sine-Gordon model: soliton versus hysteresis

Peter A. Meleshenko[1,2,*], *Akim V.* Tolkachev[1], *Mikhail E.* Semenov[1,2,3,4,**], *Alla V.* Perova[3], *Andrey I.* Barsukov[3], and *Alexander F.* Klinskikh[1]

[1]Voronezh State University, Universitetskaya sq. 1, 394006, Voronezh, Russia
[2]Zhukovsky–Gagarin Air Force Academy, Starykh Bolshevikov st. 54 "A", 394064, Voronezh, Russia
[3]Voronezh State Technical University, XX-letiya Oktyabrya st. 84, 394006 Voronezh, Russia
[4]Geophysical Survey of Russia Academy of Sciences, Lenina av. 189, 249035 Obninsk, Russia

Abstract. In this work we consider a nonlinear dynamical system which is a set of nonlinear oscillators coupled by springs with hysteretic blocks (hysteretic sine-Gordon model). The hysteretic nonlinearity is modeled using the phenomenological Bouc-Wen model). The wave processes (solitonic solutions) in such a system are investigated taking into account the hysteretic nature of the coupling.

1 Introduction

The theory of oscillations, which studies oscillations occurring in various systems, is an intensively developing field of modern mathematics and physics [1–4]. The main models of the theory of oscillations are the linear and nonlinear oscillators, rotators, RLC circuit, etc. These are used in modeling of physical processes in various real-life systems.

The standard model of wave processes is a finite and infinite chain of coupled (interacting) oscillators. Such chains are often used in radio engineering as filters that allocate or suppress signals with frequencies lying in a certain band. From the fundamental point of view, chains of oscillators are used as models of solid media with oscillations and waves with various properties [5–7]. Wave phenomena are widespread in nature: waves on the surface of a fluid, sound waves in a gas, compression-expansion waves in a solid, vibrations of a string and membrane, electromagnetic waves, etc.

In addition to nonlinear oscillations, there are also nonlinear waves the behaviour of which is described by nonlinear partial differential equations. Within the framework of the theory of nonlinear waves there exist the standard models, similar to the reference models in the theory of oscillations, namely, simple waves, shock waves, as well as the solitary waves (solitons), that play significant roles in the theory of nonlinear processes. One of the basic models for studying the nonlinear processes is the well-known sine-Gordon model (a simple mechanical realization of which is a chain of nonlinear oscillators linked by coil springs) [8].

Another example of a strong nonlinear system playing a significant role in modern researches is hysteresis. Hysteretic behavior is typical both for the characteristics of substances (e.g., ferroelectrics, ferromagnetics, piezoelectrics etc.), and for the dynamics of many mechanical systems (backlash, stop, etc.). In the mechanical systems hysteretic nonlinearities arise due to an aging of the material and must be taken into account at the modeling level for adequate description of the corresponding mechanical systems.

The purpose of this paper is to investigate the dynamics of oscillatory system taking into account the hysteretic coupling conditions between individual links of such a system. We consider the discrete hysteretic sine-Gordon model (or modified sine-Gordon model corresponding to the case when the links between pendulums contain hysteresis nonlinearities. On the basis of numerical modeling, the dynamics of soliton-like solutions in such a system is studied.

2 Bouc-Wen model

Dependencies of hysteretic type are determined by input-output correspondences, when the output depends not only on the instantaneous value of the input, but also on its behavior in the preceding moments of time (memory effect, or the hereditarity property).

Mathematical models of hysteresis-type phenomena are rather diverse and include both design models (backlash, stop, non-ideal relay, as well as their continuous analogues, namely the Ishlinskii and Preisach models [9]), and phenomenological models (S-converter, Duhem model, Bouc-Wen model etc. [10–13]). In this paper, we focus on the phenomenological approach based on the Bouc-Wen model for describing the hysteresis nonlinearity [13–15]. Let us recall some basic ideas of this model.

*e-mail: melechp@yandex.ru
**e-mail: mkl150@mail.ru

Let us consider the equation of motion of a single-degree-of-freedom system:

$$\mu\ddot{u}(t) + F(u, z) = f(t), \qquad (1)$$

where μ is the mass, $u(t)$ is the displacement, $F(u, z)$ is the restoring force and $f(t)$ is the excitation force (the overdot corresponds to the derivative with respect to time). Following the Bouc-Wen approach the restoring force is presented as (the corresponding function depends on the input and output states)

$$F(u, z) = \alpha k u(t) + (1 - \alpha) k z(t). \qquad (2)$$

From this relation it follows that the restoring force $F(u, z)$ can be divided into elastic and hysteretic parts, where k is the yielding stiffness, α is the ratio of post-yield to pre-yield (elastic) stiffnesses and $z(t)$ is the non-dimensional hysteretic parameter that obeys the following nonlinear differential equation with zero initial condition ($z(0) = 0$):

$$\dot{z}(t) = \left[A - |z(t)|^n (\beta + \operatorname{sign}(z(t)\dot{u}(t))\gamma)\right] \dot{u}(t), \qquad (3)$$

where, A, β, γ and n are non-dimensional parameters controlling the behavior of the model and sign$(\cdot)$ is the standard signum-function. Parameters β and γ control the size and shape of the hysteretic loop. Thus, such a multi-parameter model describes wide class of hysteretic systems [13–15].

3 Discrete hysteretic sine-Gordon model

The most well-known and well-studied equations in mathematical physics are equations describing the propagation of waves in a linear medium. For a nonlinear medium with hysteretic properties, there are no ready-made methods for solution of such equations.

One of the interesting results of the analysis of wave propagation processes in nonlinear media is the existence of soliton solutions – solitary waves behaving like particles. One of the models that has a soliton solution is the sine-Gordon system. In the mechanical language this system can be presented as a chain of nonlinear pendulums with elastic torsion-tied links. Let us note here that this system has many applications, including the propagation of crystal defects and domains in ferromagnetic and ferroelectric materials, the propagation of splay waves on biological (lipid) membranes, one-dimensional model of elementary particles and propagation of magnetic flux quanta in the long Josephson junction [8].

In what follows we consider a mechanical system with hysteretic links. The physical model of such a system is shown in figure 1. It is a chain of identical pendulums strung on a string and connected by springs [16]. Pendulums oscillate transversely to the direction of the chain. The principal feature of the mechanical system under consideration is that the backlash-type hysteretic nonlinearity [17] is included in the link between two neighboring pendulums. This system is a modification of the classical mechanical sine-Gordon system and can be called hysteretic sine-Gordon system.

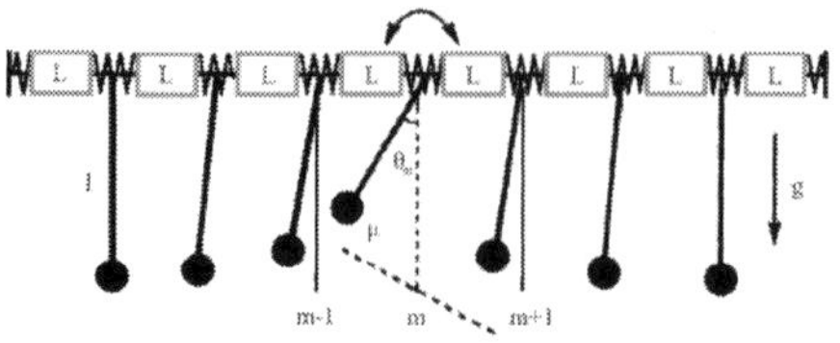

Figure 1. Discrete hysteretic sine-Gordon system

Let μ be the mass of the pendulum, μl^2 is the moment of inertia, l is the length, and κ is the torsion constant of the spring. When the deviation of the pendulum with number m from the equilibrium point by an angle θ_m takes place, the gravitational force moment $-\mu g l \sin\theta_m$ acts on the pendulum alongside the torsional moment acting on the side of adjacent springs $-\kappa(\theta_m - \theta_{m-1}) + \kappa(\theta_{m+1} - \theta_m)$. Since the hysteretic nonlinearity is included in the system, the equation of motion can be presented as:

$$\begin{cases} \mu l^2 \ddot{\theta}_m = -\mu g l \sin\theta_m + \omega_m^{left} + \omega_m^{right}, \\ \omega_m^{left} = L[\omega_m^{left}(t_0); y_m^{left}(t_0)] y_m^{left}(t), \\ y_m^{left} = -\kappa(\theta_m - \theta_{m-1}), \\ \omega_m^{right} = L[\omega_m^{right}(t_0); y_m^{right}(t_0)] y_m^{right}(t), \\ y_m^{right} = \kappa(\theta_{m+1} - \theta_m), \end{cases} \qquad (4)$$

where the time-dependent outputs $\omega_m^{left}, \omega_m^{right}$ and inputs y_m^{left}, y_m^{right} (these inputs are the corresponding moments affecting single pendulum from the left and right sides relative to neighbor pendula respectively) are the corresponding outputs and inputs for the physically realizable converter $L[\cdot]$ in the frame of Krasnosel'skii and Pokrovskii approach [9], and $\omega_m^{\cdot\cdot}(t_0), y_m^{\cdot\cdot}(t_0)$ are the corresponding initial states (output and input, respectively) of the converter.

4 Numerical simulation

It is known that the operator interpretation of the hysteretic nonlinearity (as presented in system (4) implies the non-smoothness of the corresponding operator. Therefore, in our numerical simulation we use the approach to hysteresis based on the Bouc-Wen model. In this case, the discrete

 https://doi.org/10.1051/matecconf/201824101027

hysteretic sine-Gordon system takes the following form:

$$\begin{cases} \mu l^2 \ddot{\theta}_m = -\mu g l \sin\theta_m - \alpha\kappa(\theta_m - \theta_{m-1}) - (1-\alpha)\kappa z_m^{left} + \\ \qquad + \alpha\kappa(\theta_{m+1} - \theta_m) + (1-\alpha)\kappa z_m^{right}, \\ \dot{z}_m^{left} = \{A - |z_m^{left}|^n [\beta + \gamma \mathrm{sign}(z_m^{left} \dot{u}_1(t))]\} \dot{u}_1(t), \\ \dot{z}_m^{right} = \{A - |z_m^{right}|^n [\beta + \gamma \mathrm{sign}(z_m^{right} (\dot{u}_2(t))]\} \dot{u}_2(t), \\ u_1(t) = \theta_m - \theta_{m-1}, \; u_2(t) = \theta_{m+1} - \theta_m. \end{cases} \tag{5}$$

We performed the numerical simulation of the dynamics of the mechanical system described by equations (5) using the 4-th order Runge-Kutta method. The chain has a finite length $m = 100$ (we recall that we consider a discrete system), and its ends are fixed. The initial conditions for pendulums

$$\theta_1(t_0), \dot{\theta}_1(t_0), \ldots, \theta_{m-1}(t_0), \dot{\theta}_{m-1}(t_0), \theta_m(t_0), \dot{\theta}_m(t_0)$$

generate a family of solitonic solutions moving with different velocities along the (m, t)-plane with reflection at the ends of the chain ($m = 1, m = 100$). For the parameters of the hysterestic blocks formalized by means of the Bouc-Wen model, $\dot{z}_m^{left}(t_0)$, $\dot{z}_m^{right}(t_0)$, the initial conditions are zero by default.

In order to study the influence of hysteresis bonds in the system, we consider the case in which the vibrations of 25th ($\theta_{25}(t_0) = 2\pi, \dot{\theta}_{25}(t_0) = 0$) and 75th ($\theta_{75}(t_0) = \pi, \dot{\theta}_{75}(t_0) = 0$) pendulums are excited with the corresponding initial conditions. Under these initial conditions, the oscillations of the corresponding components are excited in the chain (fig. 2 (top panel)). In the case when the hysteresis in the links is taken into account (fig. 2 (bottom panel)), spatial localization of oscillations is observed.

Let us consider in more detail the evolution of the states of the components of the chain $(\theta(t), \dot{\theta}(t))$ in the neighborhood of 25th pendulum. Figure 3 shows the phase portraits for 24th, 25th, and 26s pendulums, respectively together with corresponding hysteretic loops (such loops are obtained as a numerical solution to equation 3 of the Bouc-Wen model). As follows from this figure, in the absence of hysteretic links ($\alpha = 1$) the dynamics of pendulums demonstrates a complex oscillatory structure. However, in the presence of hysteresis in the links ($\alpha = 0.5, \beta = 0.1, \gamma = 0.9$), the dynamics in the neighborhood of the 25th pendulum is regularized and the stable limit cycle can be seen.

Also, we investigated the influence of the hysteretic blocks in the links between pendulums by using the methods of spectral analysis. We performed the Fourier transform for the 25th pendulum in the presence of hysteretic block ($\alpha = 0.5, \beta = 0.1, \gamma = 0.9$) and without ($\alpha = 1$) it. The corresponding results are shown in figure 4.

As it follows from the results presented in this figure, the oscillation spectrum changes after inclusion of hysteretic bonds. Thus we can conclude that the hysteresis in such a system plays a role of a "filter" that quenches frequencies corresponding to small-amplitude oscillations and releases the main frequency.

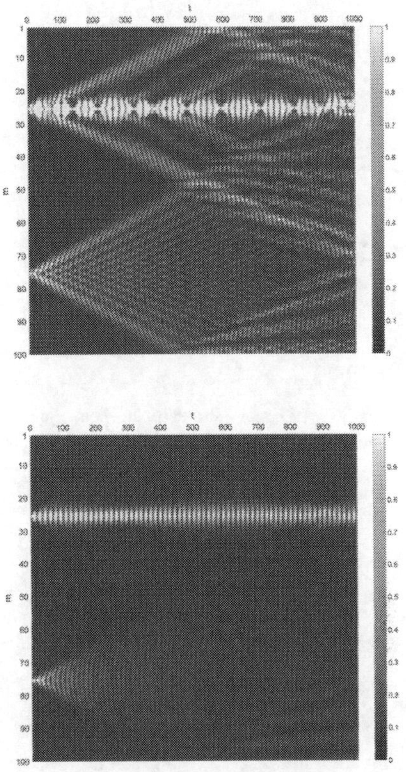

Figure 2. Simulation of the dynamics of localized (soliton-like) oscillations of pendulums in a chain without hysteresis ($\alpha = 1$, top panel) and with hysteresis ($\alpha = 0.75, \beta = 0.1, \gamma = 0.9$, bottom panel) in the links

5 Conclusions

In this paper we studied the dynamics of an oscillatory system with many degrees of freedom under conditions of hysteretic blocks in the coupling between the individual parts of the system. The system under consideration can be classified as a discrete hysteretic sine-Gordon model (the connections between the pendulums contain a hysteretic nonlinearity). The hysteretic nonlinearity was formalized by means of the Bouc-Wen model which allows a fairly simple numerical realization of the solution to the system of equations corresponding to the mechanical system under consideration. On the basis of numerical simulations, the dynamics of the solitonic solution for this system was studied taking into account the hysteretic nature of the coupling. The results of numerical simulation demonstrate the regularizing role of hysteresis bonds in the character of oscillatory motions in this mechanical system. The filtering properties of hysteretic bonds are inferred from the spectral analysis of the oscillatory motions of individual components of the system.

This work is supported by the RFBR (Grants 16-08-00312-a, 17-01-00251-a, and 18-08-00053-a)

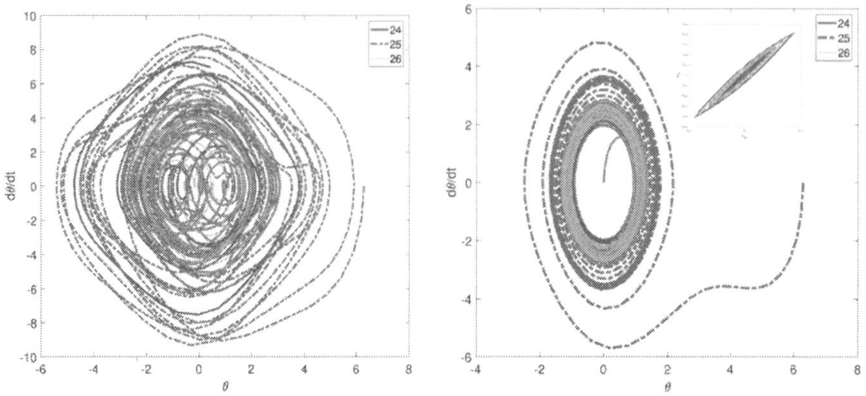

Figure 3. Phase portraits of 24th, 25th, 26s pendulums without hysteresis (top panel) and with hysteresis (bottom panel) in the links. The inset (bottom panel) shows the corresponding hysteretic loop obtained as a solution to eq. (3).

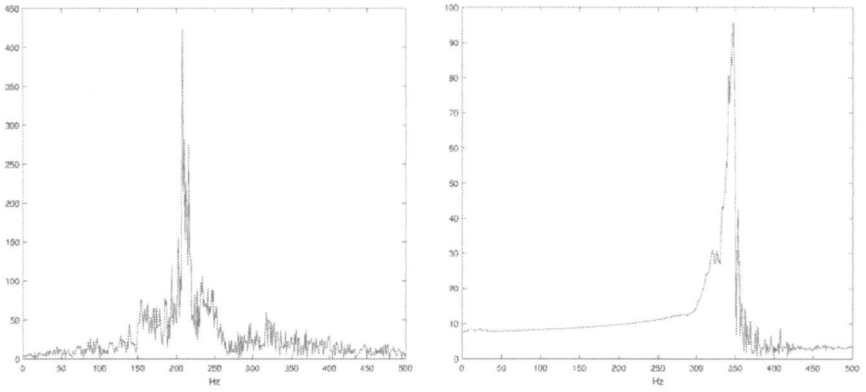

Figure 4. The oscillation spectrum of the 25th pendulum without hysteresis (top panel) and with hysteresis (bottom panel) in the link

References

[1] Lichtenberg A J, Livi R, Pettini M, Ruffo S, Lecture Notes in Physics, **728** 21–121 (2008)

[2] Torre C G, Foundations of Wave Phenomena **19** 27–38 (2014)

[3] Sieber J, Kalmar-Nagy T, Phys. Rev. E **84** 016227(1–6) (2011)

[4] Semenov M E, Solovyov A M, Popov M A, Meleshenko P A, Arch. Appl. Mech., **8**7 DOI: https://doi.org/10.1007/s00419-017-1323-0 (2017)

[5] Scott A C, *Active and nonlinear wave propagation in electronics* (Wiley-Interscience, New York, 1970) 326 p

[6] Trubetskov D I, Rozhnev A G, *Lineynyye kolebaniya i volny* (Fizmatlit, Moscow, 2001) 419 p

[7] Haken H, *Quantum field theory of solids: an introduction* (North-Holland, Amsterdam, New York, 1976) 330 p

[8] Scott A C, *Nonlinear Science. Emergence and Dynamics of Coherent Structures* (Oxford, Oxford University Prees, 1999) 474 p

[9] Krasnosel'skii M A, Pokrovskii A V, *Systems with Hysteresis* (Berlin, Springer-Verlag, 1989) 410 p

[10] Bouc R, Proceedings of the Fourth Conference on Nonlinear Oscillations (Prague, Czechoslovakia) 315—321 (1967)

[11] Bouc R, Acustica **21** 16—25 (1971)

[12] Wen Y K, Journal of the Engineering Mechanics Division **102** (2) 246—263 (1976)

[13] Ikhouane F, Rodellar J, Nonlinear Dynamics, **42** 63–78 (2005)

[14] Charalampakis A E, Archive of Applied Mechanics **85** 549–555 (2015)

[15] Solovyov A M, Semenov M E, Meleshenko P A, Barsukov A I, Procedia Engineering **201** 549–555 (2017)

[16] Scott A C, Am. J. Phys **37** 52–61 (1969)

[17] Solovyov A M, Semenov M E, Meleshenko P A, Reshetova O O, Popov M A, Kabulova E G Procedia Engineering **201** 578–583 (2017)

AUTHOR INDEX